T0205893

A Concise Introduction to Thermodynamics for Physicists

This introductory textbook provides a synthetic overview of the laws and formal aspects of thermodynamics and was designed for undergraduate students in physics, and in the physical sciences. Language and notation have been kept as simple as possible throughout the text.

While this is a self-contained text on thermodynamics (i.e., focused on macroscopic physics), emphasis is placed on the microscopic underlying model to facilitate the understanding of key concepts such as entropy, and motivate a future course on statistical physics.

This book will equip the reader with an understanding of the scope of this discipline and of its applications to a variety of physical systems.

Throughout the text readers are continuously challenged with conceptual questions that prompt reflection and facilitate the understanding of subtle issues. Each chapter ends by presenting worked problems to support and motivate self-study, in addition to a series of proposed exercises whose solutions are available as supplementary material.

Features

- Pedagogically designed, including illustrations, keyword definitions, highlights, summaries of key ideas and concepts, and boxes with additional topics that complement the materials presented in the main text.

- Presents active reading strategies, such as conceptual problems, discussion questions, worked examples with comments, end of chapter problems, and further reading to stimulate engagement with the text.

- Guides the reader with ease through a difficult subject by providing extra help whenever needed to overcome the more demanding technical and conceptual aspects.

Patrícia Faísca is Assistant Professor (with habilitation) in the Department of Physics, Faculty of Sciences, at the University of Lisbon, and principal investigator at the Biosystems and Integrative Sciences Institute. She received a PhD in Physics in 2002 from the University of Warwick as part of the Gulbenkian PhD Program in Biology and Medicine. She has a broad interdisciplinary education covering Physics, Biology and Mathematics. Her fields of interest include biological physics, thermodynamics, and statistical physics.

A Concise Introduction to Thermodynamics for Physicists

Patrícia Faísca

CRC Press
Taylor & Francis Group
Boca Raton London New York

CRC Press is an imprint of the
Taylor & Francis Group, an **informa** business

First edition published 2023
by CRC Press
6000 Broken Sound Parkway NW, Suite 300, Boca Raton, FL 33487-2742

and by CRC Press
4 Park Square, Milton Park, Abingdon, Oxon, OX14 4RN

CRC Press is an imprint of Taylor & Francis Group, LLC

ISBN: 978-0-367-55084-4 (hbk)
ISBN: 978-0-367-54641-0 (pbk)
ISBN: 978-1-003-09192-9 (ebk)

DOI: 10.1201/9781003091929

Publisher's note: This book has been prepared from camera-ready copy provided by the authors.

Typeset in Nimbus Roman
by KnowledgeWorks Global Ltd.

Dedication

*To those
who led me into Science*

Contents

SECTION I The Laws of Thermodynamics

SECTION II The Structure of Thermodynamics

SECTION III Applications

Preface

The idea to write a textbook on thermodynamics came to my mind after teaching this subject during the last eight years at the University of Lisbon. My target audience includes second year undergraduate students in Physics, Physics Engineering, and Biophysics and Biomedical Engineering. All of them have undertaken a first year of freshman Physics, being familiarised with calculus, Newtonian mechanics, and electromagnetism. Some of them will not be taught other courses on thermal physics, while others will enrol in statistical physics in their third year. Therefore, this book is an introductory text on thermodynamics, which was designed for undergraduate students in Physics and in the physical sciences.

Thermodynamics is one of the more general physical theories, with a remarkably broad scope of applicability. It is becoming an ever more interdisciplinary subject with its concepts and ideas being exported from its traditional arenas into economics, biology, and ecology. The book is a first encounter with thermodynamics and provides a synthetic overview of its laws and formal aspects. Its contents were selected to be taught in a one semester course, but it can also be used for self-study. The book is concise, and as sophisticated as possible at an introductory level. The language and mathematical notation adopted throughout the text is simple and consistent with that commonly adopted in other courses; this is essential in a first encounter with thermodynamics to avoid making it more difficult than necessary. The tools of calculus for thermodynamics (exact differentials, Legendre transforms, and homogeneous functions) are integrated in the main text as text boxes, but may be skipped if the reader is already familiarised with these topics.

This textbook presents a phenomenological, physically motivated theory of thermodynamics that smoothly and gradually unfolds into a formal and generalised theory. It is divided into three parts: Part I addresses the fundamentals and laws of thermodynamics, Part II focuses on its structure and formal aspects, and Part III explores selected applications, namely, phase transitions, magnetic systems, and thermal radiation.

While this is a self-contained book on thermodynamics (i.e. focused on macroscopic physics), emphasis is also placed on the microscopic underlying model. This facilitates the understanding of key concepts (e.g. entropy) by reducing the level of abstraction, and motivates a future course on statistical physics. Therefore, the ideal gas is used both as a thermodynamic system and a model system throughout the book, which also contains a short introduction to the kinetic theory of gases.

Nowadays, many students find it difficult to concentrate. Therefore, all presented materials are delivered in small batches. The book uses active reading strategies (including conceptual problems which are responded within the main text) to stimulate engagement with the subject through active, critical though, and raise awareness for tricky issues. Since many people, especially undergraduate students, learn from examples and from problem solving, each chapter terminates with two or three worked problems and a selection of proposed problems, whose solutions are offered online as supplementary materials.

My expectation is that by using this book, the reader will understand the fundamental principles of thermodynamics, and be able to apply it to the study of different physical systems.

I am very grateful to my collaborators and friends Antonio Rey (Universidad Complutense de Madrid), Raffaello Potestio (Università di Trento), and Rui Travasso (Universiade de Coimbra) who have carefully revised many chapters. Their helpful comments have greatly contributed to improve the clarity of the text.

I would like to deeply thank my department colleagues Margarida Cruz and Ana Nunes for their enthusiasm regarding this book and their willingness to contribute the last two chapters on magnetic systems and thermal radiation, respectively.

Finally, I am particularly pleased to acknowledge my students who have helped me to evolve as a teacher and, perhaps more importantly, with whom I have been deepening my understanding of thermodynamics.

Patrícia Faísca
Lisboa, February 2, 2022

Contributors

Ana Nunes
University of Lisbon
Lisbon, Portugal

Margarida Cruz
University of Lisbon
Lisbon, Portugal

Section I

The Laws of Thermodynamics

1 Thermodynamics Key Concepts

This chapter introduces key concepts of thermal physics and thermodynamics. Focus is placed on the notion of thermodynamic equilibrium, equilibrium fluctuations, and types of thermodynamic processes. The ideal gas is presented as a thermodynamic system as well as a model system. The concept of internal energy is discussed. Temperature is presented as an empirical property, and the zeroth law of thermodynamics is stated. The chapter ends with a brief introduction to the kinetic theory of gases, and the ideal gas equation of state is derived in this context.

1.1 INTRODUCTION

Thermal physics comprises the study of thermodynamics, kinetic theory of gases, and statistical physics. This book is focused on thermodynamics, but it will make a small detour to briefly introduce the kinetic theory of gases.

Thermodynamics was developed in the 19th century in the context of the industrial revolution. By then it was necessary to build efficient steam engines, a type of heat engine in which hot steam, usually supplied by a boiler, expands under pressure, and part of the heat is converted into work to generate motion. From an applied point of view, thermodynamics can be viewed as a branch of science and engineering whose scope is that of understanding the processes associated with energy conversion and energy transfer. However, the scope of thermodynamics is remarkably broader. Indeed, thermodynamics is a branch of Physics that studies macroscopic systems, which are systems comprising a *very large* number of components.

The system formed by the Earth and the Moon, both massive objects, is likely identified as a mechanical system. By using Newton's law of gravitation, it is possible to write the equations of motion, solve them analytically, and obtain the trajectories. If, instead of two celestial bodies, one considers three (or more), there is no longer an analytical solution. The system is still considered a mechanical system, and given accurate values for the initial positions and velocities, one can solve the equations of motion numerically and make accurate predictions for the trajectories.

If instead of a group of celestial bodies, one considers a gas inside a small box with a volume of 1 cm^3, this is likely no longer viewed as a mechanical system. Indeed, this is an example of a thermodynamic system. In this case, the number of constituent particles (atoms or molecules) is so large, that even if we had extraordinarily powerful computers to integrate the equations of motion, we would still have the problem of accurately determining a prohibitively large number of initial positions and velocities. Fortunately, for such large systems, it is possible to attain a new level of simplification and perform accurate calculations and experimental measurements. Indeed, in thermodynamics the number of particles N forming a system is

DOI: 10.1201/9781003091929-1

so large that the system's description is limited to a certain number of the so-called **thermodynamic properties**, i.e., measurable macroscopic physical properties such as the pressure (P), temperature (T), volume (V), and density $(\rho = N/V)$ of a fluid (Figure 1.1 A), without worrying about the microscopic details of the system (Figure 1.1 B). Thermodynamic properties can be also designated by **state variables** or **thermodynamic observables**.

As it will become clear as our story unfolds, thermodynamics offers a rigorous mathematical formulation on the relation between thermodynamic properties that are used to describe the *equilibrium* state of macroscopic systems, as well as the experimental methods used to measure them. It applies to systems that are sufficiently large so that equilibrium *fluctuations* can be ignored. The kinetic theory of gases, on the other hand, seeks to determine the properties of gases using the probability distributions associated with the movement of each individual particle.

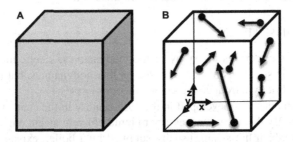

Figure 1.1 Relation between macroscopic state (A) and microscopic state (B) of a system. For a gas composed of N particles, a macroscopic state (or macrostate) is a description of a thermodynamic system based on measurements of macroscopic properties such as P, T, V, and ρ. On the other hand, a microscopic state (or microstate) is defined by the ensemble of $6N$ coordinates, which are the three position coordinates (x_i, y_i, z_i) and the three momentum coordinates $(p_{x,i}, p_{y,i}, p_{z,i})$ of each particle.

At this point it is necessary to answer the following questions:

1. What is the meaning of equilibrium and equilibrium fluctuations?
2. What is the exact meaning of a very large number of components?

1.2   THERMODYNAMIC EQUILIBRIUM

In order to understand the meaning of thermodynamic equilibrium, let us start by looking at Figure 1.2, which represents the time evolution (i.e. the value observed at time t) of some thermodynamic property X, as well as the moving average of that property. We see that after some specific time τ, termed the **relaxation time**, the moving average converges to a well-defined mean value $\overline{X}$, and $X(t)$ rapidly deviates from $\overline{X}$ by an amount $\Delta X = X(t) - \overline{X}$ that is either positive or negative. We say that X fluctuates around $\overline{X}$ and ΔX represents the **fluctuation**. A state of **thermodynamic equilibrium** is attained when the moving average becomes constant.

The random instantaneous deviations from the property's mean value observed in thermodynamic equilibrium are termed **equilibrium fluctuations**. Because of these fluctuations, which result from the fact that the system has an underlying microscopic structure (Figure 1.1 B), one cannot say that X is strictly constant in equilibrium.

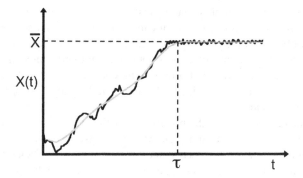

Figure 1.2   Thermodynamic equilibrium. When a system attains a state of thermodynamic equilibrium after time τ, the average value $\overline{X}$ of any thermodynamic property X becomes well defined, and $X(t)$, the instantaneous value of X, rapidly fluctuates around $\overline{X}$.

The size of equilibrium fluctuations is quantified by the so-called **relative fluctuation**, $(\overline{\Delta X^2})^{\frac{1}{2}}/\overline{X}$, with $(\overline{\Delta X^2})^{\frac{1}{2}}$ standing for the root-mean-square deviation. In consequence of a fundamental result of statistics, this quantity tends to zero as $1/\sqrt{N}$, when N, the number of independent parts of the system, tends to infinity (e.g. for a gas of N particles, the total energy is essentially the sum of order N contributions that are approximately independent). The limit where the number of particles N in a system goes to infinity ($N \rightarrow \infty$), and the system's volume V increases in proportion to N, such that N/V is finite, is termed **thermodynamic limit**. Of course, strictly speaking, physical systems are not of infinite size. However, if N is large enough (e.g. of the order of the Avogadro number $N_A = 6.02214 \times 10^{23}$), the size of relative fluctuations is smaller than the accuracy of the measurement, and fluctuations are very difficult to detect experimentally. On average, a small volume of 1.0 cm^3 of matter in the solid or liquid phase contains up to 10^{23} particles, and up to 10^{20} particles if matter is in the gas phase.

For a macroscopic system in thermodynamic equilibrium, the equilibrium fluctuations can be neglected

---

The macroscopic states of a thermodynamic system in equilibrium (also termed **macrostates**) (Figure 1.1 A) are thus characterised by properties that do not change with time. Additionally, in equilibrium, these properties are **uniform** throughout the system. On the other hand, the properties of a thermodynamic system will change with time if the system is perturbed by an external action. Think, for example, of a gas that has been pushed towards one of the sides of the container by dislocating a piston. In this case, the density on that side of the container will be higher than elsewhere, and the system will no longer be in a state of thermodynamic equilibrium. The density will vary with time until a new state of equilibrium is reached. It is important to stress that the time necessary to reach a state of thermodynamic equilibrium will depend on the specific system under study, and for some systems it can be extraordinarily long. In the case of some glasses it can be hundreds of years. Therefore, one needs to be careful when making experimental measurements because temporal independence during the observation time is not sufficient to guarantee that the system is in a state of thermodynamic equilibrium. In practice, the criteria of equilibrium is circular: a system is in a state of thermodynamic equilibrium if its properties obey the laws of thermodynamics.

One should mention as well the possibility for the system to be in a **stationary state**. The latter is a non-equilibrium state with no macroscopic time dependence. In this case, thermodynamic properties become constant within a sufficiently small part of the system. However, there always exists an energy or particle flux passing through it. In contrast, for a system in a state of thermodynamic equilibrium there is no net flow of energy or matter, either within the system or between different systems.

1.3 THERMODYNAMIC SYSTEM

So far, we used the word system to refer to any macroscopic entity composed of a large number ($N \approx N_A$) of particles. However, in thermodynamics, system has also an operational meaning. Indeed, the experimental or analytical analysis of the system can be more or less straightforward, depending on the manner according to which it is defined.

In thermodynamics the **system** is a part of the universe that we want to study. In this sentence the word universe stands for the **thermodynamic universe**, which should not necessarily be identified with the cosmological Universe, although the latter can be studied from a thermodynamics perspective. The system is enclosed by a surface termed **boundary**, which can be real or imaginary, and whose shape and size can, or not, be well defined. The boundary separates the system from the rest of the universe, which is termed **surroundings** (Figure 1.3). Depending on the type of

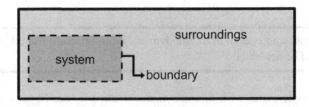

Figure 1.3 Thermodynamic system. A thermodynamic system is separated from the surroundings by a boundary. The thermodynamic universe comprises the system and the surroundings.

boundary, the surroundings can affect, or be affected, by the system. Thermodynamic systems are generally classified as:

1. **Isolated**: when neither energy nor matter can be exchanged with the surroundings.
2. **Closed**: when only energy can be exchanged with the surroundings.
3. **Open**: when both energy and matter can be exchanged with the surroundings.

Regarding the type of boundary one can consider:

1. **Diathermal** or **adiabatic**: depending on whether or not it allows for energy to be exchanged with the surroundings.
2. **Permeable** or **impermeable**: depending on whether or not it allows for matter to be exchanged with the surroundings.
3. **Movable** or **fixed**: depending on whether or not it allows for the system's volume to change.

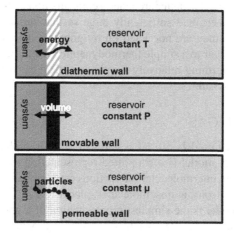

Figure 1.4   Three types of reservoirs. A system that is placed into thermal contact with a heat reservoir by means of a diathermal wall equilibrates at the reservoir's temperature. If the system is placed into mechanical contact with a volume reservoir, it equilibrates at the reservoir's pressure. Finally, if the system is placed into chemical contact with a particle's reservoir, it equilibrates at the reservoir's chemical potential.

It is important to define a specific type of surroundings that is termed **reservoir** or bath (Figure 1.4). The latter is a system whose size is so large when compared with the size of the system under study, such that the reservoir's thermodynamic properties do not change when placed into contact with the system. An idealised reservoir has an infinite size. We distinguish three types of reservoirs:

1. **Heat reservoir**: A system that can lose or gain energy while keeping a constant temperature when placed in contact with another system. When a system

is placed in thermal contact with a heat reservoir by means of a diathermal boundary, its equilibrium temperature will be identical to the reservoir's temperature.

2. **Volume reservoir**: A system that can increase or decrease its volume while keeping a constant pressure. When a system is separated from a volume reservoir by a mobile boundary, its equilibrium pressure will be identical to the reservoir's pressure.

3. **Particle reservoir**: A system that can gain or lose particles while keeping a constant *chemical potential*. When a system is placed in contact with a particle reservoir by a permeable boundary, its equilibrium chemical potential will be identical to the reservoir's chemical potential. The concept of chemical potential will be carefully addressed later on in this book. For now, it suffices to say that the chemical potential is related to the number of particles in much the same way that temperature is related to energy and pressure is related to volume.

Throughout this book, and unless otherwise stated, we will be focusing on **simple** thermodynamic systems. The latter are macroscopically homogeneous, isotropic, uncharged, chemically inert, and sufficiently large so that surface effects can be neglected. Furthermore, they are not acted on by gravitational, electric, or magnetic fields. We will often recur to a simple thermodynamic system called the ideal gas. Its simplicity allows performing relatively simple analytical calculations that illustrate the theory of thermodynamics.

1.4   THE IDEAL GAS

In Physics one creates models of real systems in order to replicate and predict the system's behaviour. A **model** is a simplified representation of the physical system under study, and of the intermolecular interactions establishing between the system's components (be it electrons, atoms, molecules, planets, etc.).

The **ideal gas model** is the simplest representation of a gas, that correctly captures a gas at low density. It represents the gas particles as point particles (i.e. a material body having mass but no spatial extent), and assumes that there are no (attractive or repulsive) physical interactions between them. The particles of an ideal gas can change energy between each other, and with the walls of the container, exclusively through **elastic collisions**, i.e., collisions that conserve momentum and energy. For an ideal gas composed of N particles in equilibrium, the pressure (P) is related with the volume (V) occupied by the gas and its temperature (T), through the **pressure equation of state**:

$$PV = Nk_BT \qquad (1.1)$$

In equation (1.1) $k_B = 1.3807 \times 10^{-23}$ JK^{-1} is the Boltzmann constant, T is the absolute (or thermodynamic) temperature, P is the absolute (or thermodynamic) pressure (measured relative to absolute zero pressure, which is the pressure of a perfect vacuum). The SI unit of temperature is the *kelvin* (K). In the Kelvin temperature scale 0 K corresponds to the absolute zero. The SI unit of pressure is the *pascal* (Pa). Perhaps, a more familiar version of the pressure equation of state is $PV = nRT$, with n being the amount of substance, whose SI units is the *mole*, and $R = 8.3145\,\mathrm{Jmol^{-1}K^{-1}}$ being the gas constant. In order to switch between the two forms if suffices to take into account that $n = N/N_A$ and $k_B = R/N_A$. The pressure equation of state is commonly known as the ideal gas equation. The ideal gas equation was stated in 1834 by Benoit Emile Clapeyron (1799–1864) and results from combining the *old* gas laws:

1. If temperature is kept constant, the pressure is proportional to the density:

$$P \propto \rho$$

   This is known as **Boyle's law**, empirically established in 1662 by Robert Boyle (1627–1691).
2. If volume is kept constant, the pressure is proportional to the temperature:

$$P \propto T$$

   This is known as **Amonton's law**, empirically established in 1702 by Guillaume Amonton (1663–1705).
3. If pressure is kept constant, the volume is proportional to the temperature T:

$$V \propto T$$

   This is (often inaccurately) named as **Charles' law**. As a matter of fact, it was established empirically in 1802 by Joseph Gay-Lussac (1778–1850).

The original formulation of the ideal gas law has, therefore, an empirical basis but it can be derived in the context of statistical mechanics and of the kinetic theory of gases, as we will show later on in this chapter.

The ideal gas equation shows that in order to characterise the equilibrium state of a closed system it is enough to specify the value of two thermodynamic properties. Indeed, if one writes the equation of state as

$$T = \frac{PV}{Nk_B},\tag{1.2}$$

with N and k_B being constants, T can be taken as a function of P and V, $T = T(P,V)$, being actually fixed by the values of P and V. If we write the equation of state such that $P = P(V,T)$, then P stays determined by V and T. Finally, if one considers $V = V(P,T)$, V stays fixed once we specify P and T.

Properties P and V in equation (1.2) are termed independent properties and T is the dependent one. Two properties are **independent** if the value of one of them can be changed without affecting the other. For a simple, single-phase fluid system with a fixed number of particles, the equilibrium state stays completely specified by two independent properties (P,V), (V,T) or (P,T). When the number of particles N is allowed to change it is necessary to consider three thermodynamics properties, and one of them must necessarily be N. For more complex systems, it is likely that more than two independent thermodynamic properties are necessary to specify the equilibrium state; in that case the equation of state is more complicated being generally expressed as $T = T(X_1,X_2,X_3,...,X_n)$. Note that a mathematician would express the functional relation between the different variables (or thermodynamic properties) using the letter f for the function and T for the property, $T = f(X_1,X_2,X_3,...,X_n)$. In Physics we simplify this notation by using the same letter for both function and variable.

It is interesting to note that the ideal gas law shows that the pressure depends on the number of particles forming the gas, but does not depend on any property of the particles (e.g. their mass). Likewise, any set of N particles sufficiently diluted always occupy the same volume V, when T and P are kept constant.

In the next section we introduce another equation of state of the ideal gas, this time for a rather important thermodynamic property called internal energy.

1.5   INTERNAL ENERGY

In Newtonian mechanics, in order to calculate the kinetic energy of a mechanical system (e.g. a body with a certain mass sliding down an inclined plane) one considers the translation of the system's centre of mass in a given coordinate axis. The potential energy results from the interaction of the system with the Earth's gravitational field. In thermodynamics one does not consider these forms of energy, which are associated with macroscopic movement, because the system under study is usually at rest. Instead, one considers the so-called internal energy, that we denote by the capital letter U.

Consider a thermodynamic system composed of N interacting particles. For the sake of simplicity, let us assume for now that the particles are atoms. In this case the total energy of the system, which we denote by E, is

$$E = \frac{1}{2}\sum_{i=1}^{N}m_i v_i^2 + \frac{1}{2}\sum_{i\neq j}u_{ij}(r),$$

where the first term is the kinetic energy of the system of particles E_k (which is the sum of the kinetic energy of all particles measured in a reference frame at rest relative to the system's centre of mass), while the second term is the potential energy of the system of particles, E_p (which is the sum of the energies associated with the total number of pairs of intermolecular interactions that establish between the system's particles). In the equation above $u_{ij}(r)$ stands for the interaction energy between particles i and j, and r represents the distance between them. Typically, for non-charged particles, the intermolecular energy potential, $u(r)$, is given by

$$u(r) = 4\varepsilon\left[\left(\frac{\sigma}{r}\right)^6 - \left(\frac{\sigma}{r}\right)^{12}\right], \tag{1.3}$$

and has the shape represented in Figure 1.5. The first term in (1.3) represents the repulsive short-range interactions, and the second term represents the attractive long-range interactions. When the distance between the particles is r_{min}, u takes a minimum value u_{min}.

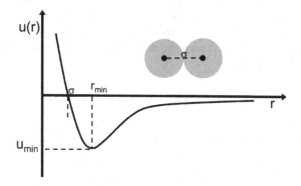

Figure 1.5 Intermolecular potential. Interaction potential between two electrically neutral particles separated by a distance r. The distance σ corresponds to the sum of the van der Waals radii of the two atoms, and r_{min} is the distance at which the interaction is optimal (i.e. corresponds to the lowest interaction energy). The interaction potential is strongly repulsive when $r < \sigma$. The attractive tail, which has a $1/r^6$ dependence in distance, results from dipole-dipole interactions between the atoms.

If energy is allowed to enter or leave the system, E will change with time such that $E = E(t)$. In thermodynamic equilibrium the **internal energy** is defined as

$$U = \langle E(t) \rangle = \frac{1}{(t_f - t_0)} \int_{t_0}^{t_f} E(t)dt, \qquad (1.4)$$

or, equivalently

$$U = \langle E_K \rangle + \langle E_p \rangle,$$

with $t_0 > \tau$. We recall that τ is the relaxation time (Figure 1.2).

If the system under study is an ideal gas, the internal energy only contains the kinetic term because there are no intermolecular interactions between the particles. For an ideal gas of N particles, a calculation based on statistical mechanics shows that the internal energy is given by

$$U = \frac{3}{2}Nk_BT \qquad (1.5)$$

The equation above is another equation of state of the ideal gas. It shows that when N is fixed the internal energy only depends on the absolute temperature T, $U = U(T)$.

In the case of a real gas, where the particles interact with each other, the energy also contains the potential energy term. Furthermore, if the particles have internal structure (as in the case of a gas is formed by diatomic particles such as O_2 molecules), one has to consider the kinetic energy of vibration, E_V (corresponding to the stretching of the chemical bond), and the kinetic energy of rotation, E_R (corresponding to the rotation of the molecule around one or more axis of rotation). Other energy terms may contribute to E, depending of the specific system under study. The internal energy associated with the disorganised, small-scale motion, is often designated as **thermal energy**.

1.6   EXTENSIVE AND INTENSIVE PROPERTIES

Consider an ideal gas formed by N particles in equilibrium enclosed in a container with volume V at absolute temperature T. Every time a particle collides with the walls of the container, a force is applied to the container's surface. The pressure P is defined as the force per unit area A. Let U be the internal energy of the gas. Now imagine that the size of the container doubles, together with the number of particles. Which thermodynamics properties keep their values in the larger system? Only the P, T, and ρ. The latter are termed **intensive** properties. An intensive property does not depend on the size (or extent) of the system; it is a scale invariant. On the other hand, an **extensive** property scales *linearly* with the system's size. Properties such as U, V, m (mass), n, and N are all examples of extensive properties; they will double their values upon doubling the size of the system. The ratio between two extensive properties is an intensive property. The molar mass M is therefore an intensive property.

Mathematically, extensive properties are **homogeneous functions of first order**, while intensive properties are **homogeneous functions of order zero**. Some properties are neither extensive nor intensive. Mathematically, these properties are still homogeneous functions, but not of degree one or zero.

HOMOGENEOUS FUNCTIONS

Let $f = f(x,y,z)$ be an homogeneous function of order k such that

$$f(\lambda x, \lambda y, \lambda z) = \lambda^k f(x,y,z), \text{ with } \lambda > 1$$

f is considered an homogeneous function of first order if $k = 1$, and an homogeneous function of order zero if $k = 0$.

Think about it...
Can you think of a property that is neither extensive nor intensive?

Answer

Consider a sphere of radius R and volume V, such that $V = (4/3)\pi R^3$. Expressed as a function of V the radius is $R(V) = (3/4\pi)^{\frac{1}{3}} V^{\frac{1}{3}}$. Thus

$$R(\lambda V) = (3/4\pi)^{\frac{1}{3}} \lambda^{\frac{1}{3}} V^{\frac{1}{3}} = \lambda^{\frac{1}{3}} R(V),$$

showing that (relative to the volume of a sphere) the radius is an homogeneous function of degree $1/3$.

1.7 THERMODYNAMIC PROCESSES

A thermodynamic process occurs whenever the system moves from one initial equilibrium macrostate i to a final equilibrium macrostate f. We already mentioned that the number of independent variables characterising a macrostate depends on the system under study. For the sake of simplicity, let us consider an ideal gas formed by a fixed number of particles. In this case the number of independent variables is two. Let us consider that one of the variables is kept fixed during the process. Now for a process to occur, one must externally perturb the system by changing one thermodynamic property. Let us denote it by X. During a thermodynamic process X will move from an initial equilibrium state where it takes the value X_i, to a final equilibrium state where it takes the value X_f, such that a finite change $\Delta X = (X_f - X_i)$ is observed. In thermodynamics we are not interested in studying how such transitions occur, nor in determining how long they take to occur, i.e., in evaluating relaxation times. Rather, we are interested in studying the measurable physical properties of equilibrium states. The designation thermodynamics is therefore somehow misleading.

It is natural to expect that an external perturbation will drive the system into a non-equilibrium state before it reaches its final equilibrium state. On the other hand, the equation of state is a mathematical relation between the system's equilibrium properties. Does this imply that the equation of state does not hold for thermodynamic processes? In other words, does it imply that processes cannot be described thermodynamically? In order to answer this question it is convenient to envisage a thermodynamic process as a continuous succession of intermediate states. Furthermore, it is also useful to classify the intermediate states as:

1. An equilibrium state of the system and the surroundings (i.e. an equilibrium state of the universe). In this case the process is termed **reversible**.
2. An equilibrium state of the system only. In this case the process is termed **quasi-static**.

The direction of a reversible process can be reversed. This means that the thermodynamic universe can be restored *spontaneously* to its original equilibrium state, leaving no residual changes elsewhere. On the other hand, a quasi-static process may be reversible, but only if it occurs in the absence of energy dissipation (e.g. due to friction, viscous damping of fluid motion). In reality there are no such processes because they result from an infinitesimal change of the system's properties, which mean that a finite change would take an infinite time to occur. However, if the change is made sufficiently slow with regard to the system's equilibrium time, the process can be considered to approach the limit of a quasi-static process and it can be studied in the context of thermodynamics. To understand these ideas better let us analyse the following thought experiments.

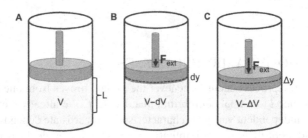

Figure 1.6   Gas compression. A gas inside a cylinder to which a frictionless piston is adjusted is in equilibrium occupying a volume V; to compress the gas, a macroscopic dislocation L of the piston should be performed (A). If the external force is infinitesimally small it will cause an infinitesimal dislocation dy and an infinite time would be necessary to cause a dislocation L of the piston (B). This process corresponds to a reversible process. A small compression of the gas is done by applying an external force that dislocates the piston by an amount of Δy (C). The successive application of small compressions would approximate the full compression process to a quasi-static process.

Consider an ideal gas formed by N particles, confined to the interior of an isolated cylinder with a piston that moves without friction. The gas is in equilibrium and occupies a volume V (Figure 1.6 A). Let us perform an infinitesimal compression of the gas by applying a downward infinitesimal force to the piston (Figure 1.6 B). This can be achieved by placing a tiny grain of sand on the top of the piston causing it to dislocate by an infinitesimal amount dy. Since the piston moves without friction, both the system (i.e. the gas inside the cylinder) and the surroundings go back spontaneously to its initial equilibrium state once grain of sand is removed, and there is no net work done in this process. In theory, one could go on, and on, by successively adding grains of sand until a macroscopic dislocation L of the piston had occurred. However, this process would take an infinite time to perform; it is an idealisation.

Now, let's consider the situation in which a downward force is applied that leads to a small macroscopic dislocation Δy (Figure 1.6 C). The density is no longer uniform because the number of particles near the piston's surface increases. This means that the system is no longer in equilibrium. Let τ be the time needed to achieve a state of thermodynamic equilibrium. After time τ the density of particles within the cylinder will be uniform again. If $\langle v_g \rangle$ is the average velocity of the particles, then it will take $\tau \approx \Delta y / \langle v_g \rangle$ for the system to equilibrate. In the context of kinetic theory of gases, we will see that $\langle v_g \rangle$ is of the order of magnitude of $10^4 - 10^5$ cm s^{-1}. On the other hand, the characteristic time of the process is given by $\tau_p \approx \Delta y/v_p$, with v_p being the velocity used to dislocate the piston in the experiment. Thus, it suffices that $\langle v_g \rangle \gg v_p$ for the system to be considered effectively in equilibrium throughout the process. In this case, we say that the process approaches the limit of a quasi-static process and thermodynamics will correctly describe the process.

In a quasi-static process it is only the system that must be in equilibrium, and a system can attain an equilibrium state after dissipating energy to the surroundings. All quasi-static processes that occur with energy dissipation are not reversible.

A reversible process is necessarily quasi-static, but a quasi-static process may not be reversible

---

In nature processes are typically **irreversible**. Irreversible processes are those whose direction cannot be reversed since it is not possible to spontaneously recover the initial state of the universe. They result from an abrupt, or sudden, change in the state of the system (e.g. a large pressure difference leading to a sudden change in volume). In irreversible processes there is always energy dissipation. Since they occur spontaneously in only one direction, irreversible processes establish the so-called *arrow of time*.

1.8 CONSTRAINTS AND PROCESSES

In the context of thermodynamics, a **constraint** is a property of the thermodynamic system which is held fixed by the observer during a thermodynamic process. Depending on the constraint, it is common to consider and perform the following types of processes:

1. **Isothermal**: the temperature of the system is constant (e.g. by placing it in thermal contact with a heat bath).
2. **Isochoric**: the volume of the system is constant (e.g. by placing the system inside a closed rigid container).
3. **Isobaric**: the pressure of the system is constant (e.g. by placing it in contact with a volume bath).
4. **Adiabatic**: there is no transfer of thermal energy between system and surroundings.

Thermodynamic processes can be represented in the so-called **state space**. This is an hyperspace whose coordinates are the several independent thermodynamic properties that describe the system's equilibrium states. In the case of a simple homogeneous fluid (like the ideal gas) with a fixed number of particles, the state space is a plane along whose two cartesian axis are plotted any of two pairs (V, T), (P, V), etc. Only equilibrium states can be represented in the state space. Any two points in the state space specify a thermodynamic change. The path connecting the two points defines a thermodynamic process.

Think about it...
What kind of process is represented in panel (A), (B), and (C) of Figure 1.7?

Answer

In panel (A) only two equilibrium states of the system could be measured corresponding to the initial and final state, respectively. The absence of a path connecting the two states means that the process is irreversible. In panel (B) a set of intermediate equilibrium points could be measured, and the distance between them is small. This process may be seen as an approximation to a quasi-static process. In panel (C) there is a continuous succession of equilibrium points linking the initial and final states of the system. The continuous line indicates that the process can be reverted. Therefore, the process in panel (C) is reversible and, by definition, it is also quasi-static.

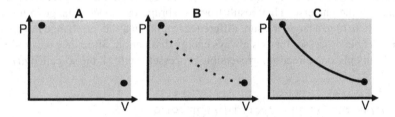

Figure 1.7   Three types of thermodynamic processes represented in the (V,P) state space.

1.9   HEAT AND HEAT CAPACITY

Heat is defined as being energy in transit from a system at higher temperature to a system at lower temperature. The *in transit* part of the definition is very important because it means that heat is not a form of energy. Instead, it is a way (or a mode) to transfer internal energy during a thermodynamic process as a result of a temperature difference; the expression **heat flow** is often used to refer to energy transferred as heat.

It is incorrect to say that a thermodynamic system contains *an amount of heat* because there is no way to identify some part of the internal energy of a thermodynamic system as being heat

---

Energy can be transferred as heat via three general mechanisms: **conduction**, also termed thermal conduction, according to which particles transfer thermal energy between each other, **convection** in which the mass motion of a fluid occurs as a result of temperature differences across that fluid (e.g. when a fluid in placed in contact with a hot source) carrying energy with it, and **radiation**, which refers to the emission and absorption of electromagnetic radiation.

When energy in transferred into a system, the system's temperature increases, and the system becomes hot. Alternatively, when energy is transferred from the system, its temperature decreases, and the system becomes cold. Let $đQ$ be the amount of energy that must be transferred into a system by heating it such that the system's temperature is raised by dT. The **heat capacity** C, is defined as:

$$đQ \equiv C dT \tag{1.6}$$

The heat capacity is then a measure of the amount of energy that is needed to increase the temperature of a system. The SI unit of heat capacity is J K^{-1}. The heat capacity often depends on temperature, $C = C(T)$. Note that the $đ$ notation in equation (1.6) is used to indicate that the infinitesimal $đQ$ is not the differential of a function. Therefore, C is not the derivative of Q with respect to T. The reason why this is so will become clear in Chapter 2.

Heat capacity is an extensive property. The so-called **specific heat**, c, is the heat capacity by mass unit $c = C/m$, and the **molar heat capacity**, C_m, is the heat capacity per mole of substance, $C_m = C/n$. Both are intensive properties.

The amount of energy transferred as heat to a thermodynamic system that changes from an initial state i whose temperature is T_i, to a final state f, whose temperature is T_f, is

$$Q_{i \to f} = \int_{T_i}^{T_f} C(T) dT. \tag{1.7}$$

When C is independent of temperature the equation above can be written as

$$Q_{i \to f} = C \Delta T,$$

with $\Delta T = T_f - T_i$.

Think about it...

Taking into account the definition of heat, does the designation heat capacity make sense?

Answer

The literal meaning of heat capacity is *capacity for holding heat*. The designation does not make sense because it is not possible to quantify the heat content of a system.

1.10 THE ZEROTH LAW OF THERMODYNAMICS

While discussing the ideal gas equation of state we said that T is the absolute or thermodynamic temperature, a quantity that is measured in the Kelvin temperature scale. However, we have not yet defined T thermodynamically. Indeed, at this stage, the notion of temperature one may have is merely sensorial. Temperature is related with the sensation of hot and cold: we say that a hot body has a higher temperature than a cold one. Empirical temperature scales, such as the Celsius and the Fahrenheit reflect our notion of hotness.

Let us consider two systems, one, termed S_l, has a low temperature T_l, and the other termed S_h, has a higher temperature T_h (Figure 1.8 A). The two bodies are placed in thermal contact such that energy can be transferred through thermal conduction between them via a diathermal wall (Figure 1.8 B). Since $T_h > T_l$, energy will be transferred spontaneously from S_h to S_l. As long as energy is being transferred, the internal energy content of the two systems will change over time, along with their temperatures. After some time we will note that:

1. The two systems exhibit the same temperature T (Figure 1.8 C), and the latter no longer changes with time.
2. The internal energy of each system does not change with time.

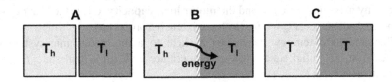

Figure 1.8   Thermal equilibrium. Two systems with different temperatures $T_h > T_l$ (A) are allowed to exchange energy via a diathermal wall (B). After some time they will equilibrate at the same temperature T (C).

Conditions 1 and 2 define a state of equilibrium termed **thermal equilibrium**. Condition 2 implies that there is not a net energy flow between the two systems: If any energy is being transferred from S_h to S_l, exactly the same amount of energy is being transferred from S_l to S_h. The process that leads to thermal equilibrium is called **thermalisation**. It is important to stress that if one starts from a situation of thermal equilibrium, the reverse process, i.e., the process of energy transfer that restores both systems to their original temperatures, will not occur spontaneously. This means that thermal processes define an *arrow of time*.

> **Think about it...**
> If systems S_l and S_h are ideal gases what is the mechanism of energy transfer that leads to thermal equilibrium?

> *Answer*
>
> Thermal equilibrium results from a process of thermal conduction driven by the elastic collisions between the gas particles.

If several systems are in thermal equilibrium with each other, their temperatures are the same. This observation motivates the statement of the zeroth law of thermodynamics.

Zeroth Law: Two systems A and B that are each one in thermal equilibrium with system C, are in thermal equilibrium with each other

---

The zeroth law justifies the use of the **thermometer**, a device that measures the temperature of a system. The thermometer is placed in thermal contact with the system and a **thermometric property** (a thermodynamic property that depends linearly on temperature, such as the pressure of a gas, the height of a mercury column, the electrical resistance of a wire, or the length of a metal rod) is measured in thermal equilibrium. A device that satisfies this request is the so-called **gas thermometer**. The latter consists of a sufficiently diluted gas (e.g. helium) enclosed in a fixed volume such as a copper bulb, which is attached to a manometer. If the system behaves as an ideal gas, the equation of state $PV = NK_BT$ should correctly capture its equilibrium behaviour. If both N and V are fixed, the relation $T(P) = CP$ holds, with C being a constant. To determine C it is necessary to calibrate the thermometer. This may done by choosing the size of the unit in such a way that a given number of units lies between two fixed points. An alternative is to use a **reference temperature**, which is the temperature of a physical process that consistently occurs at the same temperature. A commonly used reference temperature is that corresponding to the triple point (tp) of water. The triple point is the exact temperature and pressure at which liquid water, ice and water vapour coexist. The triple point temperature is defined as $T_{tp} = 273.16$ K, and the triple point pressure is $P_{tp} = 611.657$ Pa. Taking this reference temperature into account, the temperature of a gas thermometer is given by

$$T = 273.16 \left(\frac{P}{P_{tp}} \right).$$

In the gas thermometer, the thermometric property used to report the temperature of some physical system is the pressure of the gas, P. Note that this equation indicates that if the pressure is zero the temperature must be 0 K. This means that if a temperature of 0 K could be achieved, there would be no collisions with the walls of the container and, therefore, the particles should be at rest. It is important to stress that the gas thermometer will only give reliable temperature measurements as long as the gas behaves as an ideal gas. In particular, the pressure should be low and the temperature should be high. At high pressures and low temperatures, the two assumptions of the ideal gas model (namely, that the particles have no volume and establish no intermolecular interactions) are no longer valid, and the ideal gas equation breaks down.

Think about it...

For a device to work properly as a thermometer, its heat capacity should be larger or smaller than that of the system whose temperature one wishes to measure? And what about its size? Should it be larger or smaller?

Answer

The thermometer's heat capacity should be considerably smaller than that of the system. Otherwise, the system would need to transfer a large amount of energy into the thermometer, and this would change the system's original temperature. Since the heat capacity is an extensive property, the size of the thermometer should be smaller than that of the system for the same reason.

1.11   A BRIEF INTRODUCTION TO THE KINETIC THEORY OF GASES

The focus of this book is thermodynamics, the macroscopic description of physical systems. However, here we will make a small detour to discuss microscopic physics in the particular context of the **kinetic theory of gases**. The latter is a branch of thermal physics that uses Newton's equations of motion and probability distributions to explain the macroscopic properties of dilute gases. This theory was the first to provide a connection between the microscopic realm of a physical system, and the system's macroscopic properties. Many famous scientists contributed to the development of the kinetic theory of gases, but the most important contributors are James Clerk Maxwell (1831–1879) and Ludwig Boltzmann (1844–1906). It is important to mention that during the 1880s, the existence of atoms and molecules was still a matter of dispute. Indeed, it was only in 1908 that the french physicist Jean Perrin (1870–1942) was able to experimentally demonstrate the discontinuity of matter, an achievement for which he was awarded the 1926 Nobel Prize in Physics. In this context, Maxwell's and Boltzmann's insights and contributions to thermal physics are particularly remarkable.

1.11.1   VELOCITY SPACE

We consider a thermodynamic system formed by N rigid particles of a monoatomic gas without internal structure. The size of these gas particles is small when compared with the average distance separation between them, i.e., the gas is dilute. Moreover, the particles do not interact through intermolecular forces, and they move ballistically (in a straight line) until they collide elastically with each other, or with the rigid walls of the container with rigid walls. The latter has volume V, and is thermally insulated from the surroundings. The particles are given random initial velocities. After a great number of collisions among the particles, a state of thermal equilibrium will be achieved characterised by temperature T. Because there are no forces between the particles, we assume that all the positions inside the container are equality probable.

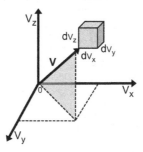

Figure 1.9   The velocity space, with an infinitesimal element volume $d\vec{v} = dv_x\, dv_y\, dv_z$.

While discussing the kinetic theory of gases, we will be working is the so-called **velocity space** (Figure 1.9), a Cartesian space whose axes are the components of velocity vector

$$\vec{v} = (v_x, v_y, v_z).$$

A point in the velocity space represents a particle, or, more specifically, the velocity vector of some particle. An element $d\vec{v} = dv_x\, dv_y\, dv_z$ of the velocity space contains a statistically significant number of particles. The velocities of the particles in the element $d\vec{v}$ are such that the x component lies between v_x and $v_x + dv_x$, while the y and z−components lie within v_y to $v_y + dv_y$ and v_z to $v_z + dv_z$, respectively.

We will follow Maxwell's footsteps (1860) to derive an expression for two probability distributions: one for the component v_i, ($i = x, y, z$) of the velocity vector, which will be represented by $g(v_i)$, and another one for the speed $v = |\vec{v}|$, which will be represented by $f(v)$. Both v_i and v are continuous variables, which means that their probability is described by a **density function**. The **probability density function** represents a continuous probability distribution.

PROBABILITY DENSITY FUNCTION

Let x be a continuous random variable. Since x is continuous it can take an infinite number of values. Therefore, instead of asking what is the probability of x taking some particular value (which is zero), we should ask what is the probability of finding x in some interval $[a,b]$,

$$P[a,b] = \int_a^b p_x(x)dx,$$

with $p_x(x)$ being the probability density function, and $p_x(x)dx$ providing the probability to find x between x and $x + dx$. Like probabilities, probability densities must be normalised. Thus

$$\int_{-\infty}^{+\infty} p_x(x)dx = 1.$$

If we wish to evaluate the mean value of x, we need to compute

$$\langle x \rangle = \int_{-\infty}^{+\infty} x p_x(x)dx.$$

Similarly, to evaluate the n-th moment

$$\langle x^n \rangle = \int_{-\infty}^{+\infty} x^n p_x(x)dx,$$

and to evaluate the n-th central moment

$$\langle (x - \langle x \rangle)^n \rangle = \int_{-\infty}^{+\infty} (x - \langle x \rangle)^n p_x(x)dx.$$

It is also important to recall that two continuous random variables x and y are said to be independent if

$$p_{x,y}(x,y) = p_x(x)p_y(y).$$

Also note that

$$p_x(x)p_y(y)dxdy$$

provides the probability that x takes values in the interval x and $x + dx$, and, simultaneously, y takes values in the the interval y and $y + dy$.

1.11.2   VELOCITY DISTRIBUTION

We define $g(v_i)$ such that

$$g(v_i)dv_i = \frac{dN_{v_i}}{N}, \tag{1.8}$$

is the **fraction of particles** whose i component of the velocity v_i lies between v_i and $v_i + dv_i$, with dN_{v_i} being the corresponding number of particles. Maxwell wanted to find the number of particles within the element volume $d\vec{v}$. To succeed in his goal, he made the supposition that, in equilibrium, the velocity distributions are isotropic (i.e. the same in the three cartesian axes):

$$Ng(v_x)dv_x = Ng(v_y)dv_y = Ng(v_z)dv_z.$$

Since the three components of the velocity vector are independent, the number of particles within the element volume $d\vec{v}$ is

$$Ng(v_x)g(v_y)g(v_z)dv_xdv_ydv_z.$$

Moreover, since all directions are equally likely, the function $\phi(\vec{v}) = g(v_x)g(v_y)g(v_z)$ must only depend on the magnitude of $\vec{v}$:

$$\phi(\vec{v}) = g(v_x)g(v_y)g(v_z) = f(v) = f\left[(v_x^2 + v_y^2 + v_z^2)^{\frac{1}{2}}\right].$$

Note that $\phi(\vec{v})dv_x dv_y dv_z$ is a probability per unit volume in velocity space. The equation above is an example of a **functional equation**, which is a type of equation in which the unknown represents a function. By inspection, a possible solution is

$$g(v_i) = Ce^{Av_i^2},$$

such that

$$f(v) = C^3 e^{Av^2}.$$

Maxwell reasoned that since the distribution must converge (when $v \to \infty$), A must be negative ($A < 0$). To determine A and C, we use the fact that since $g(v_i)$ is a density function, it must be normalised. Therefore,

$$\int_{-\infty}^{+\infty} g(v_i)dv_i = 1,$$

such that

$$C = \frac{1}{\int_{-\infty}^{+\infty} e^{-Av_i^2}dv_i}.$$

The integral in the denominator is a simple **Gaussian integral**.

GAUSSIAN INTEGRAL

A Gaussian is a function of the type $e^{-\alpha x^2}$. Gaussian integrals are very important in statistical mechanics and it is important to know how to evaluate them. Consider the integral

$$G = \int_{-\infty}^{+\infty} e^{-\alpha x^2}dx,$$

and perform a change of variables $y = x\sqrt{a}$ such that

$$G = \frac{1}{\sqrt{a}}\int_{-\infty}^{+\infty} e^{-y^2}dy.$$

To evaluate the integral above, we square it and recast the product as a two-dimensional integral:

$$G^2 = \frac{1}{\sqrt{a}}\int_{-\infty}^{+\infty} e^{-x^2}dx\frac{1}{\sqrt{a}}\int_{-\infty}^{+\infty} e^{-y^2}dy.$$

$$= \frac{1}{a}\int_{-\infty}^{+\infty}\int_{-\infty}^{+\infty} e^{-(x^2+y^2)}dxdy$$

Making a change of variables to polar coordinates

$$G^2 = \frac{2\pi}{a} \int_0^{+\infty} r e^{-r^2} dr$$
$$= \frac{\pi}{a}.$$

Thus

$$\int_{-\infty}^{+\infty} e^{-\alpha x^2} dx = \sqrt{\frac{\pi}{a}}.$$

Two general formulas for the integral of a Gaussian are:

$$\int_{-\infty}^{+\infty} x^{2n} e^{-\alpha x^2} dx = \frac{2n!}{n!2^{2n}} \sqrt{\frac{\pi}{\alpha^{2n+1}}}, \tag{1.9}$$

and

$$\int_0^{+\infty} x^{2n+1} e^{-\alpha x^2} dx = \frac{n!}{2\alpha^{n+1}}. \tag{1.10}$$

Using equation (1.9) (with $n = 0$), one gets $C = \sqrt{\frac{A}{\pi}}$, and consequently

$$g(v_i) = \sqrt{\frac{A}{\pi}} e^{-Av_i^2}. \tag{1.11}$$

Therefore

$$N\phi(\vec{v})dv_x dv_y dv_z = N\left(\frac{A}{\pi}\right)^{\frac{3}{2}} e^{-A(v_x^2+v_y^2+v_z^2)} dv_x dv_y dv_z$$

is the number of particles within the volume element $d\vec{v}$.

To succeed in achieving Maxwell's goal we still need to determine $f(v)$ and A. Let us do so by performing a change of variables to spherical coordinates:

$$\phi(v_x, v_y, v_z)dv_x\, dv_y\, dv_z = \phi(v, \theta, \varphi)v^2 \sin\theta dv d\theta d\varphi = \left(\frac{A}{\pi}\right)^{\frac{3}{2}} e^{-Av^2} v^2 \sin\theta dv d\theta d\varphi.$$

Since $\phi(v, \theta, \varphi)$ is independent of θ and φ, one can integrate ϕ over these variables

$$\int_0^{2\pi} \int_0^{\pi} \phi(v, \theta, \varphi)v^2 \sin\theta dv d\theta d\varphi = \frac{4}{\sqrt{\pi}} A^{\frac{3}{2}} v^2 e^{-Av^2} dv.$$

Thus

$$f(v)dv = \frac{4}{\sqrt{\pi}} A^{\frac{3}{2}} v^2 e^{-Av^2} dv. \tag{1.12}$$

At this point we recall that in a gas without intermolecular interactions, the energy only contains the kinetic terms. Therefore, the average energy of one gas particle is

$$\frac{1}{2}m\langle v^2\rangle.$$

In the thermodynamic limit it is expected that U as given by equation (1.4) is identical to

$$U = \frac{N}{2}m\langle v^2\rangle.$$

Moreover, we know from equation (1.5) that $U = \frac{3}{2}Nk_BT$ for a gas of non-interacting particles. It then follows that

$$\langle v^2\rangle = 3k_BT/m.$$

On the other hand, the average value $\langle v^2\rangle$, is also given by

$$\langle v^2\rangle = \int_0^{+\infty} v^2 f(v)dv = \frac{4}{\sqrt{\pi}}A^{\frac{3}{2}} \int_0^{+\infty} v^4 e^{-Av^2} dv.$$

Considering (1.9) with $n = 2$:

$$\int_0^{+\infty} v^4 e^{-Av^2} dv = \frac{1}{2}\left(\frac{3}{4}\sqrt{\frac{\pi}{A^5}}\right).$$

Thus

$$\langle v^2\rangle = 3/2A,$$

and since this should be identical to $3k_BT/m$, it comes that $A = m/2k_BT$.

By inserting A in equation (1.11) we have thus succeeded in achieving Maxwell's goal in finding an expression for the distribution of the velocity component $g(v_i)$:

$$g(v_i) = \left(\frac{m}{2\pi k_BT}\right)^{\frac{1}{2}} \exp\left(-\frac{mv_i^2}{2k_BT}\right) \tag{1.13}$$

Similarly, by inserting A in equation (1.12), we obtain an expression for $f(v)$:

$$f(v) = 4\pi\left(\frac{m}{2\pi k_BT}\right)^{\frac{3}{2}} v^2 \exp\left(-\frac{mv^2}{2k_BT}\right) \tag{1.14}$$

Equation (1.14) is designated by **Maxwell-Boltzmann (MB) distribution**. The MB distribution is represented in Figure 1.10 A. It has a long tail, and exhibits a strong dependence on the temperature, with the curve broadening, and v shifting to higher values as the temperature increases.

Think about it...
What do the shaded areas indicated in Figure 1.10 B represent?

Answer

The first, smaller area represents the fraction of particles with velocities between v_a and v_b, $\frac{N_I}{N} = \int_{v_a}^{v_b} f(v)dv$, and the larger area represents the fraction of particles with velocity larger than v_c, $\frac{N_{II}}{N} = \int_{v_c}^{+\infty} f(v)dv$.

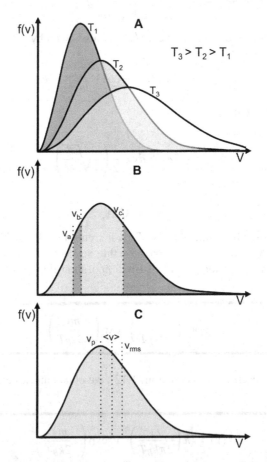

Figure 1.10 The Maxwell-Boltzmann distribution at three different temperatures (A), interpretation of the area under the MB curve (B), and the distribution at T_2 highlighting three important quantities: the most likely speed (v_p), the average speed $\langle v \rangle$, and the *rms* speed, v_{rms} (C).

Having derived the MB distribution, one can determine the **average speed of a gas particle**

$$\langle v \rangle = \int_0^{+\infty} v f(v) = \sqrt{\frac{8k_B T}{\pi m}}, \tag{1.15}$$

and the **root mean square speed** of a gas particle

$$v_{rms} = \sqrt{\langle v^2 \rangle} = \sqrt{\frac{3k_B T}{m}}. \tag{1.16}$$

The most likely speed v_p, $\langle v \rangle$ and v_{rms} are indicated in the curve represented in Figure 1.10 C. These quantities depend on the particle's mass and therefore they will be different for different gases. The average speed of a molecule of N_2, at room temperature ($T = 300$ K) is 515 ms^{-1}.

Interestingly, the experimental verification of the MB distribution was only achieved in 1955 by Miller and Kusch, who performed high-resolution measurements of the velocity distribution in beams of potassium and thallium over a range of speeds from 0.3 to 2.5 times the most likely speed, having obtained a very good agreement with the MB distribution.

1.11.3   DERIVING THE PRESSURE EQUATION OF STATE

We are now in conditions to evaluate an expression for the gas pressure by using Newton's laws of motion, and the probability distribution $\phi(\vec{v}) = g(v_x)g(v_y)g(v_z)$, with $g(v_i)$ given by (1.13).

If a particle with velocity $\vec{v} = (v_x, v_y, v_z)$ collides with a rigid surface of area A at an angle θ (Figure 1.11), its momentum in the x-direction changes according to $\Delta p_x = -2p_x$. A momentum change of equal magnitude and opposite direction is transferred to the rigid wall

$$\Delta p_x = 2p_x,$$

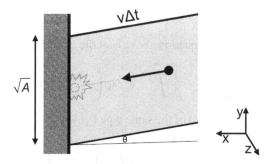

Figure 1.11   Adopted setup for the calculation of pressure. A particle with velocity $\vec{v}$ collides with a rigid wall of area A whose perpendicular makes an angle θ with $\vec{v}$. During a time interval Δt the particles colliding with with velocity $\vec{v}$ sweep a volume $Av\cos\theta\Delta t$, with $v\cos\theta = v_x$.

due to one collision.

On the other hand, the number of collisions, dn_{col}, with the surface in a time interval Δt due to particles in the volume element $d\vec{v}$ is given by

$$dn_{col} = (Av_x\Delta t) \times \left(\frac{dN_{\vec{v}}}{V}\right), \tag{1.17}$$

where the first term represents the volume swept by the particles with velocity $\vec{v}$ that collide with the surface in Δt, and the second term is the density of those particles.

If we multiply and divide (1.17) by N, we can rewrite it as

$$dn_{col} = (Av_x\Delta t) \times \left(\frac{N}{V}\right) \times \left(\frac{dN_{\vec{v}}}{N}\right), \tag{1.18}$$

with

$$\frac{dN_{\vec{v}}}{N} = \phi(\vec{v})dv_x dv_y dv_z = \left(\frac{m}{2\pi k_B T}\right)^{\frac{3}{2}} \exp\left(-\frac{m(v_x^2 + v_y^2 + v_z^2)}{2k_B T}\right) dv_x dv_y dv_z,$$

being the **fraction of particles** in the volume element $d\vec{v}$.

The total force F acting perpendicularly to the surface is the force due to *all* collisions. It is equal to the total momentum transferred to the surface during a time interval Δt divided by Δt. Since each particle transfers a momentum $2mv_x$, the total momentum transferred during Δt is equal to $2mv_x$ times the number of particles that collide with the surface during Δt. The total force F is thus obtained by integrating

$$2mv_x dn_{col}$$

in all directions of the velocity. In doing so, we note that

$$0 < v_x < +\infty$$

in the adopted setup represented of Figure 1.11, and therefore

$$F = 2mA\left(\frac{N}{V}\right)\left(\frac{m}{2\pi k_B T}\right)^{\frac{3}{2}} \int_0^{+\infty} dv_x v_x^2 \exp\left(-\frac{mv_x^2}{2k_B T}\right) \int_{-\infty}^{+\infty} dv_y \exp\left(-\frac{mv_y^2}{2k_B T}\right) \int_{-\infty}^{+\infty} dv_z \exp\left(-\frac{mv_z^2}{2k_B T}\right).$$

Note that the first integral in equation above is of the type (1.9) with $n = 1$:

$$\int_0^{+\infty} v_x^2 \exp\left(-\frac{mv_x^2}{2k_B T}\right) dv_x = \frac{1}{2}\int_{-\infty}^{+\infty} v_x^2 \exp\left(-\frac{mv_x^2}{2k_B T}\right) dv_x = \frac{\sqrt{\pi}}{4}\left(\frac{2k_B T}{m}\right)^{\frac{3}{2}}.$$

The other two integrals are also of the same type but with $n = 0$:

$$\int_{-\infty}^{+\infty} \exp\left(-\frac{mv_i^2}{2k_B T}\right) dv_i = \left(\frac{2\pi k_B T}{m}\right)^{\frac{1}{2}}.$$

Thus

$$F = 2mA\left(\frac{N}{V}\right)\left(\frac{m}{2\pi k_B T}\right)^{\frac{3}{2}} \frac{\sqrt{\pi}}{4}\left(\frac{2k_B T}{m}\right)^{\frac{3}{2}}\left(\frac{2\pi k_B T}{m}\right).$$

After simplifying the expression above, and taking into account that $P = F/A$, one finally obtains the ideal gas equation of state:

$$PV = Nk_BT.$$

Think about it...

Why did we obtain the ideal gas pressure equation of state, despite the fact that the derivation did not require the particles to be point particles?

Answer

Because we used a result from statistical physics, namely, that $U = \frac{3}{2}Nk_BT$, which is derived for an ideal gas system.

1.12 LEARNING OUTCOMES

At the end of this chapter the reader is expected to:

1. Identify the scope of thermodynamics and be able to distinguish it from other disciplines of thermal physics, namely, the kinetic theory of gases.
2. Understand the meaning of thermodynamic equilibrium and equilibrium fluctuations, relaxation time, and thermodynamic limit.
3. Know the basic concepts of thermodynamics (e.g. thermodynamic system, boundary and reservoir; open, closed and isolated system; thermodynamic properties; equation of state).
4. Know the difference between extensive and intensive thermodynamic properties and their relationship with homogeneous functions.
5. View the ideal gas as model system and list its assumptions, as well as a thermodynamic system.
6. Appreciate the importance of the ideal gas equation of state and know the meaning of independent thermodynamic properties.
7. Understand what is internal energy and why it differs from the energy of a mechanical system.
8. Be able to classify a thermodynamic process as irreversible, quasi-static, or reversible.
9. Know the meaning of constraint in relation to thermodynamic processes.
10. Understand that heat is a way to transfer energy driven by a temperature difference and not a form of energy.
11. Be able to relate temperature, heat, and heat capacity.
12. Be able to state the zeroth law of thermodynamics and identify its implications, namely, the possibility to measure temperature with thermometers.
13. Know what is thermal equilibrium and the meaning of thermalisation.
14. List the assumptions of the kinetic theory of gases.

15. Understand the derivation of the Maxwell-Boltzmann distribution and appreciate its significance.
16. Be able to derive the ideal gas equation in the context of the kinetic theory of gases.
17. Relate temperature with average speed.

1.13 WORKED PROBLEMS

PROBLEM 1.1
A cylinder whose diameter is 4 cm contains an ideal gas compressed by a piston of mass $m = 13$ kg that can move in the cylinder without friction (Figure 1.12). The cylinder and piston are immersed in a heat reservoir whose temperature can be controlled. The system is initially at equilibrium at temperature $T_i = 20°$C. The initial height of the piston above the bottom of the cylinder is $h_i = 4$ cm. The temperature of the water bath is slowly increased to a final temperature $T_f = 100°$C. Determine the final height h_f of the piston.

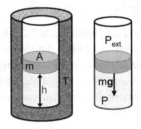

Figure 1.12   A cylinder, tapped by a frictionless piston of mass m in thermal contact with a bath at temperature T.

Solution

A slow temperature raise indicates that the process can be taken as a quasi-static process during which the gas inside the cylinder will be in equilibrium. The equilibrium condition requires that the total force acting on the piston must be null throughout the process. Assuming that the piston has area A, P is the pressure of the gas (which is the pressure inside the cylinder), and P_{ext} is the external pressure. The equilibrium condition implies that

$$PA - P_{ext}A - mg = 0.$$

Solving for P one gets

$$P = \frac{mg}{A} - P_{ext},$$

which remains constant along the isobaric process. For the initial state i of the system with $V_i = Ah_i$, the ideal gas pressure equation of state is

$$PAh_i = Nk_BT_i.$$

Noting that no particles enter or leave the system, for the final state f one has

$$PAh_f = Nk_BT_f.$$

Solving for h_f, one gets

$$h_f = \frac{h_iT_f}{T_i}.$$

Taking into account that $T_f = 393$ K, $T_i = 293$ K, and $h_i = 0.04$ m, one finally obtains $h_f = 0.05$ m.

PROBLEM 1.2

The temperature of an ideal gas in a tube of very small cross-sectional area A varies linearly from one end $(x = 0)$ to the other end $(x = L)$, according to the equation

$$T(x) = T_0 + \frac{T_L - T_0}{L}x.$$

If the volume of the tube is V, and the pressure is uniform throughout the tube, show that the equation of state for n moles of gas is given by

$$PV = nR\frac{T_L - T_0}{\ln\left(\frac{T_L}{T_0}\right)}.$$

Solution

The considered system is an example of a stationary (non-equilibrium) system. In order to solve this exercise one must consider that the system is locally (i.e. at each point x) in equilibrium. If we consider a small piece of tube located at distance x from the edge, one can ask how many moles of substance are there between x and $x + dx$. Let this amount be denoted by dn. Since the system is locally in equilibrium, the pressure equation of state holds, and one can write

$$P(x)dV(x) = dnRT(x).$$

Since the pressure is uniform, $P(x) = P$, and taking into account that $dV(x) = Adx$,

$$PAdx = dnRT(x).$$

Thus

$$dn = \frac{PA}{R}\frac{dx}{T(x)}.$$

Integrating the equation above from $x = 0$ to $x = L$:

$$n = \frac{PA}{R} \int_0^L \frac{dx}{\left(T_0 + \frac{T_L - T_0}{L}x\right)}.$$

Making the change of variables

$$u = \left(T_0 + \frac{T_L - T_0}{L}x\right),$$

one obtains

$$n = \frac{PA}{R} \frac{L}{T_L - T_0} \int_{T_0}^{T_L} \frac{du}{u}.$$

Since $V = AL$, the number of moles is finally given by

$$n = \frac{PV}{R} \frac{1}{T_L - T_0} \ln\left(\frac{T_L}{T_0}\right).$$

1.14 SUGGESTED PROBLEMS

PROBLEM 1.3
Use the ideal gas pressure equation of state to determine the density ρ of an ideal gas at standard atmospheric temperature ($T = 20°C$) and pressure ($P = 1$ atm).

PROBLEM 1.4
The mean radius of an air molecule (O_2 or N_2) is 0.15 nm. The average distance between nearest neighbour particles in the gas is 3.5×10^{-9} m at standard atmospheric pressure and room temperature. Can the air be modelled as an ideal gas?

PROBLEM 1.5
Consider the isothermal, isobaric, and isochoric processes of the ideal gas represented in Figure. 1.13. Determine the order (from smaller to larger) between T_1, T_2 and $T_3, P_1, P_2,$ and $P_3,$ and $V_1, V_2,$ and V_3.

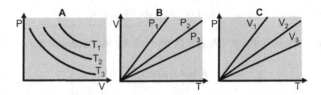

Figure 1.13   Three types of thermodynamic processes of the ideal gas.

PROBLEM 1.6

A closed cylinder has a freely moving diathermal piston separating two chambers of lengths L_1 and L_2. Chamber 1 contains 25 mg of N_2, while chamber 2 contains 40 mg of helium gas. In thermodynamic equilibrium, what is the ratio L_2/L_1? And the ratio between the number of moles of N_2, to the number of moles of He?

PROBLEM 1.7

Estimate the number of molecules in an isothermal atmosphere as a function of height.

Hint: Model the atmosphere as an ideal gas and consider a molecule at temperature T subjected to gravity.

PROBLEM 1.8

A vertical right cylinder oh height $h = 30$ cm and base area $A = 12$ cm^2 is sitting open under standard atmospheric temperature and pressure. A 5.0 kg piston is placed into the cylinder and allowed to move without friction to a final equilibrium position. Assuming the final temperature to be $T = 0°$C, what is the equilibrium height and pressure?

PROBLEM 1.9

From a thermodynamics standpoint, what is the difference between a *hot* baseball at rest and a *cold* baseball moving at high speed?

PROBLEM 1.10

Consider a sphere of radius R and volume V. Show that when expressed as a function of V, $R^3(V)$ is an homogeneous function of degree one.

PROBLEM 1.11

Classify the following processes as reversible, irreversible, or quasi-static.
a) Squeezing a plastic bottle.
b) Ice melting in a glass of hot water.
c) Pumping air into a tyre.
d) Compressing a gas with a real piston (i.e. with friction).

PROBLEM 1.12

The Celsius temperature scale is defined by setting the freezing point of water equal to $T_f = 0°$C and the boiling point of water equal to $T = 100°$C. On the other hand, the Kelvin (or absolute) temperature scale is defined by setting the freezing point of water equal to $T_f = 273.15$ K and the boiling point of water equal to $T = 373.25$ K. Determine the relation between the two temperature scales, i.e., show that $T(K) = T(°C) + 273.15$.

PROBLEM 1.13
Consider the Maxwell-Boltzmann distribution. Evaluate v_p.

PROBLEM 1.14
Use a software of your choice to evaluate the Maxwell-Boltzmann distribution of the following gases at $T = 273$ K: O_2, N_2, He, H_2, H_2O. What do you conclude?

PROBLEM 1.15
Starting from the Maxwell-Boltzmann distribution, determine $f(E_K)dE_K$, i.e., the fraction of gas particles with kinetic energy between E_K and $E_K + dE_K$.

PROBLEM 1.16
Evaluate the fraction of oxygen molecules with $199 < v < 201$ (ms^{-1}) at $T = 27°C$.

PROBLEM 1.17
In a gas, which fraction of particles have $v < v_{rms}$ and $v > v_{rms}$?

PROBLEM 1.18
Derive an expression of the number of collisions between the particles of an ideal gas and a planar surface of area A, during time interval Δt, by unit area and unit time. Express your answer as a function of the gas pressure P.

PROBLEM 1.19
A solid surface with dimensions 2.5 nm by 3.0 nm is exposed to gaseous argon at $T = 500$ K and $P = 90$ Pa. How many collisions occur with the surface during 15 seconds?

REFERENCES

1. Baierlein, R. (1999). Thermal Physics. Cambridge University Press.

2. Callen, H. B. (1960). Thermodynamics. Wiley.

3. Feynman, R. P. (2014). Exercises for the Feynman Lectures. Basic Books; New Millennium ed. Edition.

4. Girolami, G. S. (2020). A brief history of thermodynamics, as illustrated by books and people. J. Chem. Eng. Data 65 (2): 298-311.

5. Gould, H. & Tobochnik, H. (2010). Statistical and Thermal Physics: With Computer Applications. Princeton University Press.

6. James, W. S. (1929). The discovery of the gas laws. Science progress in the twentieth century 24 (93): 57-71.

7. Poirier, B. (2014). A Conceptual Guide to Thermodynamics. Wiley.

8. Robitaille, P. M., & Crothers, S. J. (2019). Intensive and extensive properties: Thermodynamic balance. Physics Essays 32 (2): 158-163.

9. Waldram, J. R. (1985). The Theory of Thermodynamics. Cambridge University Press.

2 The First Law

This chapter is dedicated to the first law of thermodynamics. It starts by defining thermodynamic work, W, and presents the first law as $\Delta U = Q + W$. To formally distinguish heat and work from thermodynamic state functions, the meaning of exact differential is presented. The first law is applied to the study of the isothermal and adiabatic compression of the ideal gas, and to analyse a cyclic process of the ideal gas.

2.1 INTRODUCTION

This chapter is dedicated to the first law of thermodynamics, often perceived as the most easy to grasp. While the zeroth law is centred on temperature, the first law is dedicated to energy, and provides a framework to better understand this important, but somehow elusive physical concept. Indeed, as we shall see, the first law represents a general statement of the law of energy conservation, which is presented for mechanical systems in the context of introductory courses on Newtonian mechanics. There, energy conservation is discussed in association with changes in kinetic and potential energy, together with their relationship to work. Thermodynamics generalises the law of energy conservation by including the effects of heat and thermodynamic work on the internal energy of a system. The first law, as we know it today, was stated by Rudolph Clausius (1822–1888), and is used extensively in the analysis of the so-called heat engines, which are devices that played a critical role in the path that lead to the formulation of the second law of thermodynamics.

2.2 THERMODYNAMIC WORK

The internal energy of a thermodynamic system can be modified through two different ways. One of them, heat, was discussed in the previous chapter. The other, work, is the focus of the present section. In thermodynamics, **work** is defined as energy in transit. Contrary to what happens when energy is transferred as heat, a temperature difference is not directly involved with an energy transfer as work. Work is energy transfer via the macroscopically observable properties of a system (e.g. the volume in the case of a fluid). Work always involves bulk movement, which microscopically leads to a net molecular flux. Heat, one the other hand, involves a direct energy transfer between the particles that constitute the system, without causing a net molecular flux.

Let us consider a system composed by an ideal gas of N particles that is confined to the interior of a cylinder, which is tapped on one of its sides by a piston of area A. The piston moves without friction. Let us assume as well that the mass of the piston is negligible ($m \approx 0$), and that a force of magnitude F_{ext} is applied to the piston creating an external pressure $P_{ext} = F_{ext}/A$. Let P be the gas pressure (i.e. the pressure inside the cylinder). If the system is in thermodynamic equilibrium, $P_{ext} = P$. If a differential

DOI: 10.1201/9781003091929-2

(or infinitesimal) pressure difference $dP = P - P_{ext}$ exists between the system and the surroundings, the piston will be reversibly dislocated by an infinitesimal distance dy. Since $F = PA$, **the reversible work done on the system** is defined as

$$\dj W \equiv -P dV \tag{2.1}$$

where we took into account that $dV = A dy$. The minus sign in (2.1) ensures that $\dj W > 0$ if the gas is compressed ($dV < 0$). On the other hand, the work done by the gas $(-\dj W)$ is positive when the gas expands ($dV > 0$). As in the definition of heat capacity (1.6), the notation $\dj$ is used to indicate that the infinitesimal $\dj W$ is not the differential of a function.

Using the ideal gas equation, one can rewrite (2.1) as

$$\dj W = -\frac{N k_B T}{V} dV. \tag{2.2}$$

Since the process is reversible, and N is constant, (2.2) can be integrated from the initial equilibrium state i to the final equilibrium state f, to obtain the total energy transferred as work:

$$W_{i \to f} = -N k_B \int_{V_i}^{V_f} \frac{T}{V} dV. \tag{2.3}$$

Think about it...

Would it be it correct to use the notation ΔW instead of $W_{i \to f}$?

Answer

The notation ΔW suggests that it is possible to measure an initial and a final amount of work in the system such that $\Delta W = W_f - W_i$. Therefore, it is not correct. The same is true in the case of heat, where one uses the notation $Q_{i \to f}$ to denote an amount of energy transferred as heat in a thermodynamic process that takes the system from the initial equilibrium state i to the final equilibrium state f.

Let us further assume that the process is isothermal (T is constant). In this case, by evaluating the integral $\int_{V_i}^{V_f} \frac{dV}{V}$ one obtains

$$W_{i \to f} = -N k_B T \ln \left(\frac{V_f}{V_i} \right). \tag{2.4}$$

According to (2.4), energy $W_{i \to f}$ is transferred *to the* system from the surroundings in the case of a compression, and energy $-W_{i \to f}$ is transferred *from the* system to the surroundings when the gas expands. In the first case one says that the system **consumes** work, while in the second case one says that it **produces** work.

If a large pressure difference $\Delta P = P - P_{ext}$ exists, a sudden change will cause the process to be irreversible due to energy dissipation, and the ideal gas equation only applies to the initial and final equilibrium states. However, if the external pressure is uniform, the work done on the system can still be defined as $đW = -P_{ext}dV$. If $P_{ext} > P$ the gas will compress ($dV < 0$). Since reversible work done on the gas is $đW = -PdV$, it is easy to see that $đW > -PdV$. This inequality also holds for an expansion of the gas ($P_{ext} < P$ and $dV > 0$), with the work done by the gas $(-đW)$ being less than the work done in a reversible expansion, $(-đW) < PdV$. As a matter of fact, the work the system does reversibly is the *maximum* work the system can produce. This can be easily understood on physical grounds because energy lost by dissipation in irreversible processes is energy that could have been used to produce work during a thermodynamic process.

Think about it...

To transfer energy reversibly as work it is necessary to perform an infinitesimal dislocation of the piston such that $dP = P - P_{ext}$. How can energy be transferred reversibly as heat?

Answer

If S_1 and S_2 are two systems in thermal contact, for energy transfer to occur reversibly, an infinitesimal temperature difference must exist between the two systems (i.e. $T_1 = T_2 + dT$) . If so, the hottest system may become the cooler by an infinitesimal reduction of its temperature, and the direction of energy transfer may be reversed.

Work and heat depend on the particular steps used to perform a thermodynamic process. We say that work and heat are **path** dependent. This can be easily illustrated in the case of work. Consider the processes represented in Figure 2.1. Both are **cyclic processes**, which start and end at the same point in the (V,P) plane. However, while cycle A is performed clockwise, cycle B is performed in a counter clockwise manner. Since $V_1 = V_4$ and $V_2 = V_3$ both cycles contain two isochoric processes for which the work is zero, $W_{1\to4} = W_{4\to1} = 0$ and $W_{2\to3} = W_{3\to2} = 0$. The remaining processes are isobaric. For a process at constant P, the total energy transferred as work from the initial state i to a final state f is

$$W_{i\to f} = -P \int_{V_i}^{V_f} dV = -P(V_f - V_i).$$

Thus the total work of cycle A is

$$W_A = W_{1\to2} + W_{3\to4}$$
$$= (P_l - P_h)(V_2 - V_1) < 0,$$

and that of cycle B is

$$W_B = W_{2\to1} + W_{4\to3}$$
$$= (P_h - P_l)(V_2 - V_1) > 0,$$

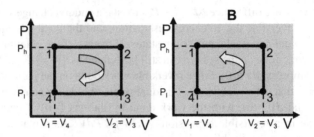

Figure 2.1   Two thermodynamic cycles in the (V,P) state space.

showing that work is indeed path dependent.

2.3 THE FIRST LAW OF THERMODYNAMICS

Having discussed energy transfer as heat and work, we are now in conditions to state the first law of thermodynamics.

First Law (system): The variation of internal energy of a closed system is given by

$$\Delta U = Q + W \tag{2.5}$$

In the equation above ΔU represents the variation of internal energy of a system as a result of a thermodynamic process that drives the system from an initial state i to a final state f, $\Delta U = U_f - U_i$, while Q and W represent energy transferred *to the system* as heat and work, respectively. An alternative convention, is to consider energy transferred as work *from the system* to the surroundings. In that case $\Delta U = Q - W$, with the minus sign indicating the work done by the system.

A process is designated as **adiabatic** when $Q = 0$. In that case $\Delta U = W$. Alternatively, if $W = 0$, $\Delta U = Q$. If the thermodynamic system is isolated, Q and W are both zero, and $\Delta U = 0$. The thermodynamic universe is itself isolated. For the thermodynamic universe the first law reads:

First Law (universe): The internal energy of the thermodynamic universe is

$$\Delta U_{universe} = \Delta U + \Delta U_{surroundings} = 0 \tag{2.6}$$

This is the most general form of the first law, which shows that energy is conserved: whenever energy is gained by the system it must be lost by the surroundings and vice-versa ($\Delta U = -\Delta U_{surroundings}$).

2.4   EXACT DIFFERENTIALS AND STATE FUNCTIONS

The first law shows that contrary to what happens with work and heat, the change in internal energy upon a process depends only of the final and initial states of the system (U_f and U_i, respectively), being independent of the path used to connect them. U is said to be a **state function** or function of state.

If a function $f(x)$ is a state function,

$$\int_{x_i}^{x_f} df = f(x_f) - f(x_i) = \Delta f, \tag{2.7}$$

which implies that

$$\oint df = 0. \tag{2.8}$$

From a mathematical point of view, for a function f to be a state function its total differential df must be **exact**. In other words, if the total differential df is exact, then f is a state function.

EXACT DIFFERENTIAL

Consider a differentiable function of two independent variables x and y, $f(x,y)$. Differentiating by parts we have

$$df = \left(\frac{\partial f}{\partial x}\right)_y dx + \left(\frac{\partial f}{\partial y}\right)_x dy.$$

The expression above is the **differential** of function f. The differential provides the value of an infinitesimal change in f as a result of infinitesimal changes in the independent variables. Likewise, it has one term for each variable, consisting of a partial derivative multiplied by the differential of the independent variable.

The differential

$$dF = Mdx + Ndy$$

is **exact**, if and only if, $dF = df$ for all x and y.

If dF is exact,

$$M = \left(\frac{\partial f}{\partial x}\right)_y \text{ and } N = \left(\frac{\partial f}{\partial y}\right)_x.$$

On the other hand

$$\left(\frac{\partial M}{\partial y}\right)_x = \frac{\partial^2 f}{\partial y \partial x} \text{ and } \left(\frac{\partial N}{\partial x}\right)_y = \frac{\partial^2 f}{\partial x \partial y}.$$

Since

$$\frac{\partial^2 f}{\partial y \partial x} = \frac{\partial^2 f}{\partial x \partial y},$$

then

$$\left(\frac{\partial M}{\partial y}\right)_x = \left(\frac{\partial N}{\partial x}\right)_y \qquad (2.9)$$

It may be shown that (2.9) is a necessary and sufficient condition for dF to be exact. For that reason it is termed the **exactness condition**.

All measurable thermodynamic properties (such as U, T, V, P) that can be used to specify the state of a thermodynamic system are state functions that have exact differentials. On the other hand, the amount of energy transferred as heat or work depends on the path that moves the system between states. Sometimes the expression **path function** is used in association with work and heat.

The reader is now in conditions to understand the formal reason behind the use of the $đ$ notation in $đQ$ and $đW$. It denotes the fact that since heat and work are not state functions, their infinitesimals cannot be expressed as exact differentials.

Think about it...

How can you use (2.9) to show that T, V, and P are state functions?

Answer

Taking $T = T(P,V)$, the differential dT is

$$dT = \left(\frac{\partial T}{\partial P}\right)_V dP + \left(\frac{\partial T}{\partial V}\right)_P dV.$$

For T to be a state function dT must be exact.
Consider the ideal gas equation $PV = nRT$.
Let

$$M = \left(\frac{\partial T}{\partial P}\right)_V = \frac{V}{nRT},$$

and

$$N = \left(\frac{\partial T}{\partial V}\right)_P = \frac{P}{nRT}.$$

It is easy to see that dT is an exact differential, because

$$\left(\frac{\partial M}{\partial V}\right)_P = \left(\frac{\partial N}{\partial P}\right)_V = \frac{1}{nRT}.$$

The same reasoning can be used to show that P and V are also state functions.

2.5   REWRITING THE FIRST LAW

Let us go back to the first law of thermodynamics. If one considers an infinitesimal change of the system (keeping N constant), the first law must be written as:

$$dU = dQ + dW \qquad (2.10)$$

If the system under study is a fluid then

$$dU = dQ - PdV. \qquad (2.11)$$

At this point it is important to clarify that the number of state functions is not limited to the ones we already encountered while discussing a fluid like the ideal gas. As stated before, thermodynamics is a very general physical theory and can be used to study any macroscopic system. Other state functions are the surface tension (γ) and the area (A) of a liquid film, the electric field ($\vec{E}$) and polarisation ($\vec{P}$) of a dielectric material, the tension ($\vec{T}$) and length (L) of a metallic (or elastic) rod, the chemical potential (μ) and the number of particles N, just to mention a few examples. Therefore, thermodynamic work is not limited to energy transferred during the expansion (or compression) of a fluid (the so-called expansion work), and also includes energy transferred by stretching films (elastic work), charging an electric system (electric work), adding or removing particles from the system (chemical work). In general, this is taken into account by writing work dW as a **generalised work**

$$dW = Xdx, \qquad (2.12)$$

or $dW = \vec{X} \cdot d\vec{x}$, for vectorial quantities, where X stands for some generalised force (an intensive property) (e.g. P, γ, $\vec{E}$, $\vec{m}$, and μ) and x is the corresponding generalised displacement (an extensive property) (e.g. V, A, $\vec{P}$, $\vec{B}$, and N). Some examples of thermodynamic work are thus: $dW = -PdV$, $dW = \gamma dA$, $dW = \vec{E} \cdot \vec{P}$, and $dW = \mu dN$. In Chapter 8 we will study in detail the specific case of magnetic work, which is work added to a magnetic system in order to magnetise it.

2.6   JOULE EXPERIMENT

From a experimental point of view, the most important demonstration of the first law of thermodynamics is the paddle-wheel experiment, based on a series of remarkably accurate experimental measurements carried out by James Prescott Joule (1818–1889), that culminated with the publication of a very famous article in 1850 entitled *The mechanical equivalent of heat*. To carry out his experiment, Joule attached weights to strings and pulleys, and connected them to a paddle-wheel that could rotate in the inside of a copper cylinder covered with adiabatic walls made of wood (Figure 2.2). Upon raising the weights of mass m to an appropriate height h, and by subsequently letting them drop slowly, the paddle-wheel rotates and stirs

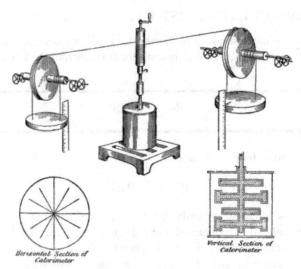

Horizontal Section of
Calorimeter

Vertical Section of
Calorimeter

Joule's Water-Churning Apparatus for Determining the
Mechanical Equivalent of Heat.

Figure 2.2   The set-up used in Joule's paddle-wheel experiment. Wikimedia commons.

the liquid (Joule used water, mercury, and sperm whale oil) of mass M placed on the inside of the cylinder. The cylinder contained fixed sections that prevented the liquid from rotating. The potential energy of the weights is converted into kinetic energy of the paddles, which is essentially converted into fluid friction that causes energy transfer as heat, with the consequent increase of the liquid's temperature by ΔT. Joule found out that it was necessary to raise the weights about 20 consecutive times to register a temperature increase of $0.5°C$–$2°C$. In this experiment, mechanical work is thus used to change the internal energy of the system. In particular, it shows that mechanical energy of the weights is converted into energy transferred as heat $(Q = C\Delta T)$ in the fluid. In the particular case of water

$$mgh = Mc_{H_2O}\Delta T,$$

which can be written as,

$$c_{H_2O} = \frac{mgh}{M\Delta T}.$$

Joule found out that 4157 J of mechanical work (the actual value is 4186 J!) were necessary to increase by $1°C$, the temperature of 1 kg of water (measured in vacuum), when the air temperature is between $12.7°C$ and $15.5°C$. This quantity, which Joule termed **the mechanical equivalent of heat,** is, in nowadays language, the specific heat of water $c_{H_2O} = 4.186$ J/g$°C$. It is also the definition of (small) calorie, i.e., 1 cal $= 4.186$ J.

> **Think about it...**
> Why is it important that the vessel used by Joule for his experiments contains fixed sections that prevent the liquid from rotating?
>
> *Answer*
> If the liquid rotates, part of the work of the gravitational force would be converted into rotational kinetic energy of the liquid, which is not internal energy.

2.7 USING THE FIRST LAW

Let us recall the definition of heat capacity (1.6). Since Q is not a state function, the heat capacity depends on the mode according to which the system is heated. In particular, the temperature increase resulting from energy transfer may occur at constant pressure or at constant volume. In the former case we define the heat capacity at constant pressure, denoted by C_P, while in the latter we define the heat capacity at constant volume, denoted by C_V. The subscripts P and V in the definition of heat capacity identify the constraint. In what follows, we will start by illustrating the importance of the first law of thermodynamics by deriving an expression for the thermodynamic definition of C_V and C_P. Subsequently, we will use the first law to study in detail the isothermal compression and the adiabatic compression of the ideal gas.

2.7.1 HEAT CAPACITY AT CONSTANT VOLUME

For a system where the number of particles N is fixed, the equilibrium state stays specified by two thermodynamics properties. To derive the thermodynamic definition of C_V, it is useful to consider the internal energy (which is a state function) as a function of T and V, $U = U(T,V)$. In this case

$$dU = \left(\frac{\partial U}{\partial T}\right)_V dT + \left(\frac{\partial U}{\partial V}\right)_T dV. \tag{2.13}$$

On the other hand, since $dQ = CdT$, according to the first law one can write

$$dU = CdT - PdV. \tag{2.14}$$

Solving (2.13) and (2.14) for CdT one gets

$$CdT = \left(\frac{\partial U}{\partial T}\right)_V dT + \left[P + \left(\frac{\partial U}{\partial V}\right)_T\right]dV.$$

Thus, the **heat capacity at constant volume** ($dV = 0$) is defined as

$$C_V \equiv \left(\frac{\partial U}{\partial T}\right)_V \qquad (2.15)$$

In the particular case of the ideal gas, $U = \frac{3}{2}Nk_BT$ and $C_V = \frac{3}{2}Nk_B$. However, in general, C_V will depend on temperature. Therefore,

$$\Delta U = Q = \int_{T_i}^{T_f} C_V(T)dT. \qquad (2.16)$$

Think about it...
Is it true that for the ideal gas $dU = C_V dT$?

Answer

By using the thermodynamic definition of C_V we can write

$$dU = C_V dT + \left(\frac{\partial U}{\partial V}\right)_T dV.$$

For the ideal gas $U = U(T)$. Consequently, the second term of the equation above vanishes. So, it is true that $dU = C_V dT$ for the ideal gas.

2.7.2 HEAT CAPACITY AT CONSTANT PRESSURE

To derive the thermodynamic definition of heat capacity at constant pressure, it is useful to consider $U = U(T,P)$. In this case

$$dU = \left(\frac{\partial U}{\partial P}\right)_T dP + \left(\frac{\partial U}{\partial T}\right)_P dT. \qquad (2.17)$$

Solving (2.14) and (2.17) for CdT one gets

$$CdT = PdV + \left(\frac{\partial U}{\partial P}\right)_T dP + \left(\frac{\partial U}{\partial T}\right)_P dT. \qquad (2.18)$$

The term dV is not useful, but one can get rid of the dependence on V by taking into account that for $V = V(T,P)$,

$$dV = \left(\frac{\partial V}{\partial P}\right)_T dP + \left(\frac{\partial V}{\partial T}\right)_P dT. \qquad (2.19)$$

Substituting dV in (2.18) by (2.19) and rearranging one obtains

$$CdT = \left[P\left(\frac{\partial V}{\partial P}\right)_T + \left(\frac{\partial U}{\partial P}\right)_T\right]dP + \left[P\left(\frac{\partial V}{\partial T}\right)_P + \left(\frac{\partial U}{\partial T}\right)_P\right]dT.$$

Finally, the **heat capacity at constant pressure** ($dP = 0$) is defined as

$$C_P \equiv \left[P\left(\frac{\partial V}{\partial T}\right)_P + \left(\frac{\partial U}{\partial T}\right)_P \right]$$
(2.20)

Think about it...
Would we obtain the same expression for C_P by considering $U = U(P,V)$?

Answer
Yes, because $V = V(P,T)$.

An important quantity that will be often used throughout this book is the so-called **adiabatic index**

$$\gamma \equiv \frac{C_P}{C_V}$$
(2.21)

Using (2.20), it is easy to see that for the ideal gas $C_P = \frac{5}{2}Nk_B$, and consequently $\gamma = \frac{5}{3}$. Note as well that for the ideal gas

$$C_P - C_V = Nk_B = nR,$$
(2.22)

which is known as **Mayer relation**.

These simple applications of the first law anticipate that the use of exact differentials and partial derivatives is a critical part of thermodynamics calculations. Therefore, in what follows we review some important and very useful properties of partial derivatives.

PROPERTIES OF PARTIAL DERIVATIVES

Let us consider a differentiable function $f(x,y,z)$, such that the relation $f(x,y,z) = c$, with c constant, holds (an example is the ideal gas pressure equation of state). In principle, this equation can be rearranged to express one variable in terms of the other two as independent variables, $x = x(y,z)$, $y = y(x,z)$, or $z = z(x,y)$. Therefore, one can write the differentials

$$dx = \left(\frac{\partial x}{\partial y}\right)_z dy + \left(\frac{\partial x}{\partial z}\right)_y dz,$$

$$dy = \left(\frac{\partial y}{\partial x}\right)_z dx + \left(\frac{\partial y}{\partial z}\right)_x dz,$$

and

$$dz = \left(\frac{\partial z}{\partial x}\right)_y dx + \left(\frac{\partial z}{\partial y}\right)_x dy.$$

Only two of the above three differentials can be independent. If one substitutes the second equation in the first, one obtains a relation between the two independent differentials dx and dz:

$$\left[\left(\frac{\partial x}{\partial y}\right)_z\left(\frac{\partial y}{\partial x}\right)_z - 1\right]dx + \left[\left(\frac{\partial x}{\partial y}\right)_z\left(\frac{\partial y}{\partial z}\right)_x + \left(\frac{\partial x}{\partial z}\right)_y\right]dz = 0.$$

By taking z constant ($dz = 0$) one obtains the **reciprocal rule**:

$$\left(\frac{\partial x}{\partial y}\right)_z = \frac{1}{\left(\frac{\partial y}{\partial x}\right)_z}. \tag{2.23}$$

By taking x constant ($dx = 0$) one obtains

$$\left(\frac{\partial x}{\partial y}\right)_z\left(\frac{\partial y}{\partial z}\right)_x = -\left(\frac{\partial x}{\partial z}\right)_y,$$

and applying the reciprocal rule to the term on the right hand side of the equation above one obtains the **reciprocity rule**:

$$\left(\frac{\partial x}{\partial y}\right)_z\left(\frac{\partial y}{\partial z}\right)_x\left(\frac{\partial z}{\partial x}\right)_y = -1. \tag{2.24}$$

Another rule of partial derivatives that is quite useful in thermodynamics calculations is the **chain rule**. To recall the chain rule, consider two single valued functions $x = x(y)$ and $y = y(z)$. Then

$$\frac{dx}{dz} = \left(\frac{\partial x}{\partial y}\right)\left(\frac{\partial y}{\partial z}\right). \tag{2.25}$$

2.7.3 ISOTHERMAL COMPRESSION AND EXPANSION OF THE IDEAL GAS

Let us consider a reversible isothermal process that takes an ideal gas of N particles from the initial state i to the final state f (Figure 2.3). For this process $dU = 0$. Indeed, $dU = C_V dT$, and $dT = 0$ because T is constant along an isotherm.

From the first law of thermodynamics $dQ = -dW = PdV$. Taking into account that both N and T are constant

$$Q_{i\to f} = \int_{V_i}^{V_f} PdV = Nk_BT\int_{v_i}^{v_f}\frac{dV}{V} = Nk_BT\ln\left(\frac{V_f}{V_i}\right). \tag{2.26}$$

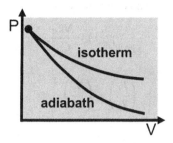

Figure 2.3 Reversible isothermal and adiabatic processes of the ideal gas. Note that the slope of the adiabath is larger than that of the isotherm.

For an isothermal compression $V_f < V_i$ and $Q_{i \to f} < 0$, which means that energy $-Q_{i \to f}$ is transferred *from the* system to the surroundings. For an expansion, $V_f > V_i$ and $Q_{i \to f} > 0$, which means that energy $Q_{i \to f}$ is transferred *to the* system from the surroundings. In the first case one says that the system **rejects** heat, while in the second case one says that the system **extracts** heat. Since $\Delta U = 0$, $W_{i \to f} = -Q_{i \to f}$. Therefore, the system **consumes** work in the compression, and **produces** work in the expansion. Finally, at constant temperature, the ideal gas pressure is directly proportional to the inverse of the volume ($P \propto \frac{1}{V}$), which means that pressure decreases in the expansion, and increases in the compression. In sum, in an isothermal compression (expansion) of the ideal gas:

1. $\Delta T = 0$ and $\Delta U = 0$
2. $PV = constant$
3. $\Delta P < 0$ (expansion)
4. $\Delta P > 0$ (compression)
5. $Q > 0$ and $W = -Q < 0$ (expansion): the system extracts heat and produces work
6. $Q < 0$ and $W = -Q > 0$ (compression): the system rejects heat and consumes work

2.7.4   ADIABATIC COMPRESSION AND EXPANSION OF THE IDEAL GAS

For an adiabatic process $dQ = 0$, and from the first law of thermodynamics $dU = dW$. Since we are considering a reversible adiabath of the ideal gas (Figure 2.3), this equation can be written as $C_V dT = -Nk_B \frac{T}{V} dV$. Integrating both sides by taking into account that both N and C_V are constant

$$\int_{T_i}^{T_f} \frac{dT}{T} = -\frac{Nk_B}{C_V} \int_{V_i}^{V_f} \frac{dV}{V}$$

$$ln\left(\frac{T_f}{T_i}\right) = \frac{Nk_B}{C_V} ln\left(\frac{V_i}{V_f}\right).$$

Now, let us go back to (2.22) and divide it by C_V:

$$(\gamma - 1) = \frac{Nk_B}{C_V}.$$

Thus

$$\left(\frac{T_f}{T_i}\right) = \left(\frac{V_i}{V_f}\right)^{(\gamma-1)},$$

or, equivalently

$$T_f V_f^{(\gamma-1)} = T_i V_i^{(\gamma-1)}. \tag{2.27}$$

Equation (2.27) is the so-called **adiabath equation**, which is normally presented as:

$$TV^{(\gamma-1)} = constant \tag{2.28}$$

An alternative form of (2.28), which can be derived from (2.27) with T given by the ideal gas equation, is:

$$PV^\gamma = constant \tag{2.29}$$

In sum, in an adiabatic compression (expansion) of the ideal gas:

1. $Q = 0$
2. $TV^{(\gamma-1)} = constant$
3. $PV^\gamma = constant$
4. $\Delta T < 0$ (expansion) and $\Delta T > 0$ (compression)

Think about it...

For an adiabatic process $Q = 0$ but the temperature decreases in the expansion and increases in the compression. Why is that?

Answer

For an adiabath of the ideal gas $dU = \mathit{d}W$, with $dU = C_V dT$. Thus, $\mathit{d}W = C_V dT$. Integrating this equation between an initial state i and a final state f, one gets

$$W_{i \to f} = C_V (T_f - T_i). \tag{2.30}$$

In the adiabatic expansion $T_f < T_i$, and therefore $W_{i \to f} < 0$ as a result of energy being transferred from the system to the surroundings. On the other hand, in the adiabatic compression $T_f > T_i$, and therefore $W_{i \to f} > 0$, as a result of energy being transferred to the system from the surroundings.

2.7.5   THERMODYNAMIC CYCLES

A thermodynamic **cycle** is a sequence of thermodynamic processes that return the system to its initial state. In principle, a cycle leaves the system unchanged, although it changes the surroundings. Since

$$\oint dU = 0,$$

$\Delta U_{cycle} = 0$. Thus, for a cyclic process the first law of thermodynamics reads

$$Q_{cycle} + W_{cycle} = 0.$$

In section 2.9, we will apply the concepts and results we have derived so far to the study of a thermodynamic cycle.

2.8   LEARNING OUTCOMES

At the end of this chapter the reader is expected to:

1. Understand the meaning of thermodynamic work and realise that work is path dependent.
2. Be able to define expansion work against a constant pressure and know that other forms of thermodynamic work exist besides expansion work.
3. Understand the relation between reversible work and maximum work.
4. State the first law of thermodynamics and understand that the internal energy of a system can be changed by transferring energy as heat and/or work.
5. Realise that the first law of thermodynamics is a generalisation of the law of energy conservation for mechanical systems.
6. Be familiarised with the Joule experiment and understand that it represents the first experimental demonstration of the first law.
7. Understand that state functions are exact differentials and know how to apply the exactness condition.
8. Use the first law of thermodynamics to derive and expression for the heat capacity at constant pressure and for the heat capacity at constant volume.
9. Use the first law of thermodynamics to quantitatively analyse an isothermal and an adiabatic expansion of the ideal gas.
10. Use the first law of thermodynamic to quantitatively analyse a thermodynamic cycle of the ideal gas.

2.9   WORKED PROBLEMS

PROBLEM 2.1
Consider the cycle represented in Figure 2.4 A, which refers to 1.00 k mol of ideal gas. The process $B \rightarrow C$ is an isothermal, $P_A = 1$ atm, $V_A = 22.4$ m^3, and $P_B = 2P_A = 2.0$ atm. Compute the total work and total heat corresponding to this cycle.

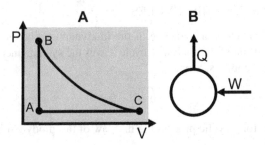

Figure 2.4   A thermodynamic cycle of the ideal gas, where $B \to C$ represents an isotherm (A), and a diagrammatic representation of the cycle (B).

Solution

The cycle is composed of three reversible processes: The isochor $A \to B$, the isotherm $B \to C$, and the isobar $C \to A$. Consider the definition of work $\dslash W = -PdV$, with P given by the ideal gas pressure equation of state

$$\dslash W = -\frac{Nk_B T}{V} dV, \tag{2.31}$$

and start by evaluating the work corresponding to each branch of the cycle.

1. **Isochoric heating $A \to B$**

   In this case $\dslash W_{A \to B} = 0$ because $dV = 0$. Thus

   $$W_{A \to B} = 0.$$

2. **Isothermal expansion $B \to C$**

   Integrating (2.31) with constant T, and taking into account that $V_B = V_A$:

   $$W_{B \to C} = -Nk_B T_B \int_{V_B}^{V_C} \frac{dV}{V} = Nk_B T_B \ln\left(\frac{V_A}{V_C}\right).$$

3. **Isobaric compression $C \to A$**

   Integrating (2.31) with constant P:

   $$W_{C \to A} = -P_A \int_{V_C}^{V_A} dV = P_A(V_C - V_A).$$

To evaluate T_B, T_C, and V_C let us consider the ideal gas pressure equation of state:

$$T_B = \frac{P_B V_A}{Nk_B} = \frac{2P_A V_A}{Nk_B} = 2T_A,$$

and

$$V_C = \frac{Nk_B T_C}{P_C} = \frac{Nk_B T_B}{P_A} = 2V_A.$$

Thus,

$$T_C = \frac{P_C V_C}{N k_B} = \frac{P_A 2 V_A}{N k_B} = 2 T_A.$$

In sum:

1. $W_{A \to B} = 0$
2. $W_{B \to C} = 2 P_A V_A \ln 0.5$
3. $W_{C \to A} = P_A V_A$

Finally, the work corresponding to one cycle is

$$W_{cycle} = W_{A \to B} + W_{B \to C} - W_{C \to A}$$
$$= P_A V_A (2 \ln 0.5 + 1).$$

Now, we evaluate the heat corresponding to each branch of the cycle.

1. **Isochoric heating $A \to B$**
   Since $W_{A \to B} = 0$, from the first law it follows that $dQ = C_V dT$. Thus

   $$Q_{A \to B} = C_V \int_{T_A}^{T_B} dT = C_V (T_B - T_A) = C_V T_A.$$

2. **Isothermal expansion $B \to C$**
   In this case $dU = 0$ because $dT = 0$. From the first law it follows that $dQ_{B \to C} = -dW_{B \to C}$. Thus
   $$Q_{B \to C} = -W_{B \to C} = -2 P_A V_A \ln 0.5.$$

3. **Isobaric compression $C \to A$**
   Using the first law $dQ = dU - dW$. Thus

   $$Q_{C \to A} = C_V \int_{T_C}^{T_A} dT - W_{C \to A}$$
   $$= C_V (T_A - T_C) - P_A V_A$$
   $$= -C_V T_A - P_A V_A$$

Finally, the heat corresponding to one cycle is

$$Q_{cycle} = Q_{A \to B} + Q_{B \to C} + Q_{C \to A}$$
$$= C_V T_A - 2 P_A V_A \ln 0.5 - C_V T_A - P_A V_A$$
$$= -P_A V_A (2 \ln 0.5 + 1),$$

For this cycle the first law of thermodynamics reads $-Q + W = 0$. In Figure 2.4 B we show a diagrammatic representation of the cycle. It is important to note that (unless otherwise stated) Q and W reported in the schematics always represent unsigned quantities (i.e. magnitudes), with the arrows indicating the direction of energy transfer.

PROBLEM 2.2
Two bodies have heat capacities C_1 and C_2, which are independent of temperature, and initial temperatures T_1 and T_2. They are placed in thermal contact. Show that their final temperature is given by $T_f = (C_1 T_1 + C_2 T_2)/(C_1 + C_2)$.

Solution

Considering that system formed by the two bodies is an isolated system, the total internal energy must be conserved

$$\Delta U = \Delta U_1 + \Delta U_2 = 0.$$

On the other hand, from the the first law of thermodynamics $\Delta U = Q$ because there is no volume work in this process. Thus

$$\Delta U_1 = \int_{T_1}^{T_f} C_1 dT = C_1(T_f - T_1),$$

and

$$\Delta U_2 = \int_{T_2}^{T_f} C_1 dT = C_2(T_f - T_2),$$

where we used the fact that C_1 and C_2 are both independent of temperature. Since $\Delta U_1 = -\Delta U_2$, the result follows.

2.10   SUGGESTED PROBLEMS

PROBLEM 2.3
Consider the differentials

$$dg = \left(x^2 + y^2\right) dx + 2xy dy,$$

and

$$dh = \left(x^2 + y^2\right) dx - 2xy dy.$$

a) Show that dg is exact and dh is non-exact.
b) Integrate dg and dh along the following paths:
Path A: step1 $x = 0 \rightarrow 3, y = 0 (dy = 0)$, step 2 $x = 3 (dx = 0), y = 0 \rightarrow 1$
Path B: step 1 $x = 0 \rightarrow 4, y = 0 (dy = 0)$, step 2 $x = 4 (dx = 0), y = 0 \rightarrow 1$, step 3
$x = 4 \rightarrow 3, y = 1 (dy = 0)$.
What do you conclude?

PROBLEM 2.4
Consider the ideal gas pressure equation of state and take

$$P = P(T, V).$$

Show that

$$\left(\frac{\partial T}{\partial P}\right)_V \left(\frac{\partial P}{\partial V}\right)_T \left(\frac{\partial V}{\partial T}\right)_P = -1.$$

PROBLEM 2.5
Let
$$f(T) = PV.$$

Show that
a)
$$\left(\frac{\partial P}{\partial T}\right)_V = \frac{1}{V}\frac{df}{dT}$$

b)
$$\left(\frac{\partial V}{\partial T}\right)_P = \frac{1}{P}\frac{df}{dT}$$

PROBLEM 2.6
Consider $U = U(T,V)$ **and show that**

$$C_V = -\left(\frac{\partial U}{\partial V}\right)_T \left(\frac{\partial V}{\partial T}\right)_U.$$

PROBLEM 2.7
Consider an adiabath and an isotherm starting at the same initial state. Show that the slope of the adiabath is higher than that of the isotherm.

PROBLEM 2.8
Consider the cycle represented in **Figure 2.5**, which refers to the ideal gas. The process $B \to C$ is an adiabath. Compute the total work and total heat, and represent the cycle in a diagram.

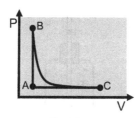

Figure 2.5   A thermodynamic cycle of the ideal gas, where $B \to C$ represents an adiabath.

PROBLEM 2.9
A thick walled insulated chamber contains n moles of helium at high pressure P_i. It is connected through a valve with a large, almost empty container of helium at constant pressure P_0, very nearly atmospheric. The valve is

opened slightly and the helium flows slowly and adiabatically into the container until the pressures on the two sides of the valve are equal. Assuming the helium behaves like an ideal gas with constant heat capacities, show that

$$T_f = T_i \left(\frac{P_f}{P_i}\right)^{\frac{(\gamma-1)}{\gamma}}.$$

PROBLEM 2.10

An ideal gas is initially confined to volume V_1 in the interior of a container with total volume $V_1 + V_2$, which is surrounded by adiabatic walls as shown in Figure 2.6. Volume V_2 is initially under vacuum. The partition that separates V_1 from V_2 is rapidly removed, the gas expands and eventually occupies volume $V_1 + V_2$. If T is the initial temperature of the gas, what will be its final temperature?

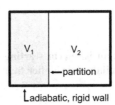

Figure 2.6   An ideal gas confined to volume V_1. Volume V_2 is under vacuum. The container is surrounded by an adiabatic wall.

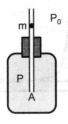

Figure 2.7   An experimental setup used to measure the adiabatic index.

PROBLEM 2.11

In the Ruchardt method to determine the adiabatic index γ, a ball of mass m is placed inside the interior of a tube with transversal section of area A, which is connected to a gas container of volume V (Figure 2.7). The gas pressure inside the container is slightly higher than the atmospheric pressure P_0,

due to force done by the ball,

$$P = P_0 + \frac{mg}{A}.$$

Show that if the ball is slightly dislocated downwards, the resulting movement will be of the simple harmonic type with period given by

$$\tau = 2\pi\sqrt{\frac{mV}{\gamma P A^2}}.$$

PROBLEM 2.12
Two systems, S_1 and S_2, with heat capacities C_1 and C_2 are placed into thermal and end up having a final temperature T_f. If T_1 is the initial temperature of system S_1, what is the initial temperature of system S_2?

REFERENCES

1. Blundell, S. J. & Blundell K. J. (2009). Concepts in Thermal Physics. Oxford University Press.

2. Callen, H. B. (1960). Thermodynamics. Wiley.

3. Young, J. (2015). Heat, work and subtle fluids: A commentary on Joule (1850) 'On the mechanical equivalent of heat'. Phil. Trans. R. Soc. A 373: 20140348.

4. Mandl, F. (1971). Statistical Physics. Wiley.

5. Pippard, A. B. (1966). Elements of Classical Thermodynamics. Cambridge University Press.

6. Reiss, H. (1996). Methods of Thermodynamics. Dover.

3 The Second Law

This chapter is dedicated to the formulation of the second law. While the zeroth law is focused on temperature and the first law on energy, the second law is about a novel state function called entropy. The chapter starts with the analysis of the Carnot cycle and of the Carnot engine. It then uses these concepts to prove the logical equivalence of the Kelvin and Clausius statements of the second law. Subsequently, Carnot's theorem is demonstrated, and entropy change for reversible processes is defined. It moves on by proving Clausius's theorem, which is then used to establish the principle of maximum entropy for isolated systems. Entropy change is calculated for different thermodynamic processes. Finally, the statistical meaning of entropy, and the relation of entropy with disorder is discussed.

3.1 INTRODUCTION

Thermodynamics was developed in the 19th century in the wake of the industrial revolution (1760–1840). The establishment of the second law, in particular, is tightly associated with the necessity to build efficient heat engines, devices that produce maximum work out of heat. A rather famous heat engine, which is considered a major driving force of the industrial revolution, is the steam engine designed by James Watt (1736–1819). Steam engines basically consisted of a combustion chamber (the *heat source*), water (the *working substance*), and a condenser (the *heat sink*). Unfortunately, steam engines were not particularly efficient.

The birth of the second law starts with the seminal work of Sadi Carnot (1796–1832), *Reflections on the Motive Power of Fire* (1824), where he recognised that in order to design efficient heat engines, it would be necessary to understand the physical principles underlying their functioning. The way he approached the problem, led him to be the first to understand that work produced by an heat engine strictly requires heat to flow from a hot body to a cold one through the working substance. In other words, he understood that engines that produce work without exchanging heat between two bodies are forbidden, and he conceived a theoretical heat engine, termed Carnot engine, which is the most efficient of all heat engines. The Carnot engine sets an upper bound for the efficiency of any heat engine designed for practical purposes.

The work of Carnot, and also that of Joule on the relation between work and heat, established the grounds for Rudolph Clausius (1822–1888) to formulate the first law of thermodynamics, and make the very first statement of the second law (1850). Later, Clausius formulated the second law mathematically by introducing a novel state function named entropy (1865). Contrary to the first law, the second law of thermodynamics has a reputation for being markedly difficult, and in part this is due to its connection with entropy, whose microscopic meaning was only established in1872 by Ludwig Boltzmann (1844–1906) in the context of statistical mechanics.

DOI: 10.1201/9781003091929-3

This chapter is dedicated to the second law of thermodynamics, which is one of the most important laws of Physics. It is the law that explains *why things happen*. Indeed, while the first law determines that all processes that conserve internal energy are possible, the second law selects those that will happen spontaneously. We start by unfolding the establishment of the second law with the analysis of the Carnot engine.

3.2 THE CARNOT ENGINE

The Carnot engine (or machine) is a theoretical heat engine envisioned by Sadi Carnot that operates a cyclic process termed **Carnot cycle**. The Carnot cycle comprises a sequence of four reversible processes of a thermodynamic system, namely, an isothermal expansion, an adiabatic expansion, an isothermal compression, and an adiabatic compression (Figure 3.1). When discussing heat engines the thermodynamic system is designated by **working substance** (after Carnot), or **working body** (after Clausius). For the isothermal expansion, the working substance is placed into thermal contact with a heat reservoir at temperature T_h, while for the isothermal compression the heat reservoir is at temperature $T_l < T_h$ (h and l stand for *high* and *low*, respectively). The heat reservoirs also have special designations. The one at temperature T_h is named **heat source**, and that at temperature T_l is the **heat sink**. Consider, for simplicity, that the working substance is an ideal gas formed by N particles. Since we already studied in detail the isotherm (section 2.7.3) and the adiabath (section 2.7.4) of the ideal gas, we can readily summarise what happens in each process that comprises the Carnot cycle (Figure 3.1):

1. **Isothermal expansion $A \to B$**
   Since $Q_{A \to B} > 0$, the system extracts heat

   $$Q_h \equiv Q_{A \to B} = N k_B T_h \ln\left(\frac{V_B}{V_A}\right) \tag{3.1}$$

   from the heat source. Since $W_{A \to B} = -Q_{A \to B} < 0$ the system produces work $-W_{A \to B}$.

2. **Adiabatic expansion $B \to C$**
   In this case $Q_{B \to C} = 0$, and since

   $$W_{B \to C} = C_V (T_l - T_h) < 0,$$

   the system produces work $-W_{B \to C}$.

3. **Isothermal compression $C \to D$**
   Since $Q_{C \to D} < 0$, the system rejects heat

   $$Q_l \equiv -Q_{C \to D} = N k_B T_l \ln\left(\frac{V_C}{V_D}\right) \tag{3.2}$$

   to the heat sink. Since $W_{C \to D} = -Q_{C \to D} > 0$, the system consumes work $W_{C \to D}$.

4. **Adiabatic compression $D \to A$**
   In this case $Q_{D \to A} = 0$, and since

   $$W_{D \to A} = C_V (T_h - T_l) > 0,$$

   system consumes work $W_{D \to A}$.

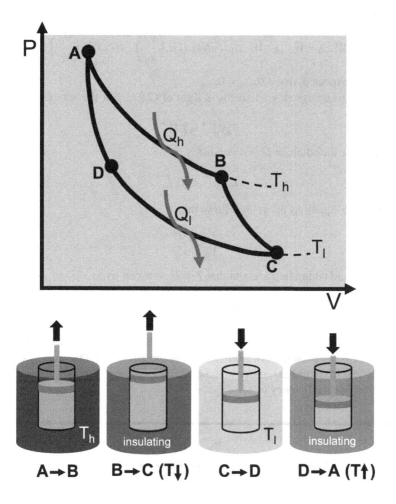

Figure 3.1 The Carnot cycle represented in the (V,P) state plane, and a schematics of the four processes emphasising the two heat reservoirs at temperatures T_h and T_l. Note that Q_h and Q_l represent unsigned quantities. This representation of the Carnot cycle is due to Émile Clapeyron (1799–1864), and was originally presented in an article published in 1834.

The heat corresponding to one cycle is thus

$$Q_{cycle} = Q_{A \to B} + Q_{C \to D} = N k_B T_h \ln\left(\frac{V_B}{V_A}\right) + N k_B T_l \ln\left(\frac{V_D}{V_C}\right).$$

Since the work of the two adiabaths cancels out

$$W_{B \to C} + W_{D \to A} = 0,$$

the work corresponding to one cycle is

$$W_{cycle} = W_{A \to B} + W_{C \to D} = -Nk_B T_h \ln\left(\frac{V_B}{V_A}\right) - Nk_B T_l \ln\left(\frac{V_D}{V_C}\right),$$

showing, as expected, that $\Delta U_{cycle} = 0$.

Now, let us analyse the adiabaths in light of (2.28). For the adiabath $B \to C$ the relation

$$T_h V_B^{\gamma-1} = T_l V_C^{\gamma-1}$$

holds, and for the adiabath $D \to A$ one has

$$T_h V_A^{\gamma-1} = T_l V_D^{\gamma-1}.$$

From the two equations above, it follows that

$$\frac{V_B}{V_A} = \frac{V_C}{V_D} \tag{3.3}$$

Using (3.3) and taking into account that $T_l < T_h$, one can write

$$W_{cycle} = Nk_B \ln\left(\frac{V_C}{V_D}\right)(T_l - T_h) < 0.$$

Thus, after one cycle, the system produces work $W = -W_{cycle}$. Since the goal of an heat engine is to produce work from heat (more precisely, from the energy Q_h transferred as heat from the heat source to the working substance), it makes sense to define the heat engine's **efficiency** as

$$\eta_{heat} = \frac{W}{Q_h} \tag{3.4}$$

Considering that

$$W = Q_h - Q_l,$$

the last equation can be written as

$$\eta_{heat} = 1 - \frac{Q_l}{Q_h} \tag{3.5}$$

Taking into account (3.1) and (3.2) it is easy to see that

$$\frac{Q_h}{Q_l} = \frac{T_h}{T_l} \tag{3.6}$$

Using (3.6) we can rewrite (3.5) as

$$\eta_{Carnot} = 1 - \frac{T_l}{T_h} \qquad (3.7)$$

Equation (3.5) shows that although the Carnot engine may be the most efficient heat engine operating between two heat reservoirs, it is not able to completely convert the energy transferred from the heat source into work. Indeed, according to (3.7) for the efficiency of a Carnot engine to be 1, either the heat source should be at infinite absolute temperature ($T_h = \infty$), or the heat sink at absolute zero ($T_l = 0$ K), which are both impossible.

Later in this chapter (section 3.5), we will use the second law of thermodynamics to demonstrate that the Carnot engine is the most efficient of all heat engines, and that all reversible engines operating between a heat source and a heat sink have the same efficiency. This last result has profound consequences. In particular, it implies that either (3.6) or (3.7) can be used to establish a temperature scale which is independent of the thermometric properties of the working substance. Either (3.6) or (3.7) thus determine the absolute, or **thermodynamic temperature** scale, named the Kelvin temperature scale in honour of Lord Kelvin who firstly established this remarkable result.

3.3 THE INVERTED CARNOT CYCLE

Since the Carnot cycle comprises four reversible processes it can run backwards. We can leave the adiabaths aside because $Q = 0$, and, as in the Carnot cycle, their combined work is zero. That leaves us with the isothermal processes. An analysis similar to the one carried out for the Carnot cycle shows that in the isotherm $B \rightarrow A$ the system rejects heat

$$Q_h \equiv -Q_{B \rightarrow A} = N k_B T_h \ln\left(\frac{V_B}{V_A}\right), \qquad (3.8)$$

to the heat source, and consumes work $W_{B \rightarrow A} = -Q_{B \rightarrow A} > 0$. On the other hand, in the isotherm $D \rightarrow C$, the system extracts heat

$$Q_l \equiv Q_{D \rightarrow C} = N k_B T_l \ln\left(\frac{V_C}{V_D}\right), \qquad (3.9)$$

from the heat sink, and produces work $-W_{D \rightarrow C} = Q_{D \rightarrow C}$.

The total energy transferred as work after one cycle is thus

$$W_{cycle} = W_{B \rightarrow A} + W_{D \rightarrow C} = N k_B \ln\left(\frac{V_C}{V_D}\right)(T_h - T_l) > 0, \qquad (3.10)$$

showing that after one cycle, the system consumes work $W = W_{cycle}$.

There are two engines that operate on the basis of an inverted Carnot cycle. One is the **refrigerator**, and the other is the **heat pump**. Their efficiency is defined differently because they have different purposes. In the case of the refrigerator, the goal is to extract energy from the cold body to keep its temperature low. Thus, for the refrigerator

$$\eta_{refrigerator} = \frac{Q_l}{W} \qquad (3.11)$$

Using (3.9) and (3.10), the efficiency of a refrigerator based on an inverted Carnot cycle can be written as

$$\eta_{refrigerator} = \frac{T_l}{T_h - T_l}, \qquad (3.12)$$

which can be larger than one.

An heat pump, on the other hand, transfers energy from a cold body to a hot body, keeping the latter's temperature high. Thus, for the heat pump

$$\eta_{pump} = \frac{Q_h}{W} \qquad (3.13)$$

Using (3.8) and (3.10) the efficiency of an heat pump based on an inverted Carnot cycle can be written as

$$\eta_{pump} = \frac{T_h}{T_h - T_l}, \qquad (3.14)$$

which is always larger than one.

We finish this section by showing a diagram of the Carnot engine (Figure 3.2 A), and of the inverted Carnot engine (Figure 3.2 B). We will used them extensively in the following sections. In this simplified representation, the black bars represent the heat reservoirs, and the circle represents a heat engine that has undergone one cycle. In the Carnot engine heat is extracted from the heat source, the system produces work W, and heat is rejected to the heat sink. In the inverted Carnot engine, on the other hand, the system consumes work W, to extract heat from the cold reservoir, and reject it into the hot reservoir. Note that W, Q_l, and Q_h represent unsigned quantities. Thus the first law of thermodynamics reads

$$Q_h - W - Q_l = 0,$$

for the Carnot engine, and

$$-Q_h + W + Q_l = 0,$$

for the inverted Carnot engine.

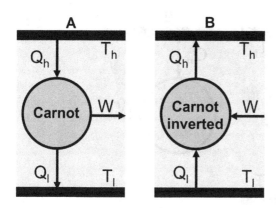

gure 3.2   The Carnot engine (A), and the inverted Carnot engine (B).

4   THE SECOND LAW OF THERMODYNAMICS

ter the publication of Joule's article that establishes the relation between work and
at, Rudolph Clausius formulated the first law of thermodynamics as we know it
day (2.5) and stated, for the first time, the second law of thermodynamics (1850).
ie year later, Lord Kelvin (1824–1907) stated the second law in an alternative way.

Second law (Clausius' statement): A process whose only result is to extract
heat from a heat reservoir and reject heat to an hotter reservoir is impossible

Second law (Kelvins' statement): A process whose only result is to extract heat
from a heat reservoir and produce work is impossible

---

must be emphasised that these statements were established on the basis of exhaus-
e empirical observations. They are verbally different but, as we will see shortly,
·y are logically equivalent. The expression *whose only result* is very important,
cause it means that once the process occurred the system is left unchanged. We
eady know that a thermodynamic process where the system is left unchanged is
·yclic process. Clausius and Kelvins' statements of the second law can be repre-
ated by using the diagrams shown in Figure 3.3 A and Figure 3.3 B, respectively.
prove their logical equivalence we will consider the following propositions:

Proposition 1: Clausius' statement implies Kelvins' statement.
Proposition 2: Kelvins' statement implies Clausius' statement.

both propositions are true, the two statements are logically equivalent. We will
ve that both propositions are true by **contraposition**. Thus, we will show that a

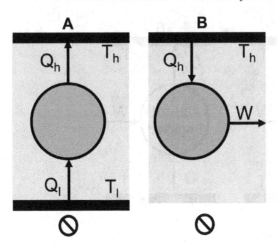

Figure 3.3   A heat engine that violates Clausius' statement (A) is not possible, and a heat engine that violates Kelvins' statement (B) is not possible.

violation of Clausius' statement leads to a violation of Kelvins' statement, and that a violation of Kelvins' statement leads to a violation of Clausius' statement.

Proof of proposition1. Let us assume that an engine that violates Clausius' statement exists, and name it Clausius violator. The latter extracts heat Q'_l from the heat sink and rejects heat Q'_h into the heat source. The Clausius violator can be combined with a Carnot engine as shown in Figure 3.4 A. Assume without loss of generality that $Q'_l = Q_l$. If not, consider two positive integers N and N' such that the ratio N/N' approximates Q'_l/Q_l with arbitrary precision. Adjust the cycles of the two engines taken as the two components of the combined system by taking N' cycles of the Clausius violator and N cycles of the Carnot engine. The combined system (Figure 3.5 B) will satisfy $Q'_l = Q_l$ within the chosen precision. By applying the first law to one cycle of the combined system ($\Delta U_{cycle} = 0$) one gets

$$Q_h - Q'_h - W = 0.$$

Thus, we conclude that the combined system only extracts heat $(Q_h - Q'_h)$ from the heat source, and produces work W, violating Kelvin's statement of the second law.

Proof of proposition 2. We assume that an engine that violates Kelvin's statement exists, and name it Kelvin violator. The latter extracts heat Q'_h from the heat source and produces work W'. The Kelvin violator can be combined with an inverted Carnot engine as shown in Figure 3.5 A. We adjust the work produced by the Kelvin violator, and the work consumed by the inverted Carnot engine such that $W = W'$ for the combined system (Figure 3.5 B). By applying the first law to one cycle of the combined system ($\Delta U_{cycle} = 0$) one gets

$$Q'_h + Q_l - Q_h = 0,$$

and therefore, the combined system only extracts heat Q_l from the reservoir at T_l, and rejects heat $(Q_h - Q'_h)$ into the reservoir at $T_h > T_l$, which violates Clausius' statement of the second law.

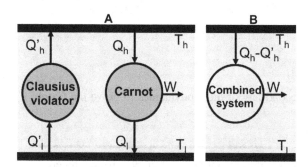

Figure 3.4   Proof of proposition 1. We assume that an engine that violates Clausius statement exists and combine it with a Carnot engine (A). The combined system only extracts heat from the reservoir at high temperature (T_h) and produces work, which violates Kelvins' statement (B).

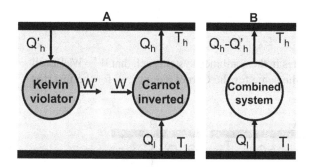

Figure 3.5   Proof of proposition 2. We assume that an engine that violates Kelvin's statement exists and combine it with an inverted Carnot engine (A). The combined system only extracts heat from the reservoir at low temperature (T_l) and rejects it into the reservoir at higher temperature $(T_h > T_l)$, which violates Clausius' statement (B).

Think about it...
Consider an isothermal process of the ideal gas. Does it violate Kelvins' statement of the second law of thermodynamics?

Answer

No because although $Q = W$, the complete conversion of heat into work is not the
only final result of the process. There is also a change in the volume of the gas,
which increases in the expansion and decreases in the compression.

3.5   CARNOT THEOREM

In this section we will use the second law of thermodynamics to prove the Carnot
theorem.

Carnot theorem: Of all the heat engines operating between two given temperatures, none is more efficient than a Carnot engine

---

To demonstrate this theorem by contraposition let us consider two engines operating
between two heat reservoirs as before. One of the engines is an inverted Carnot engine, and the other is an hypothetical irreversible heat engine named *Irr* (Figure 3.6
A). Let us assume that the efficiency of *Irr* is higher than the efficiency of the Carnot
engine $\eta_{Irr} > \eta_{Carnot}$:

$$\frac{W'}{Q'_h} > \frac{W}{Q_h}.$$

Adjust the cycles of the two engines in the combined system such that $W' = W$. Recall
that the unsigned quantity W is the same for the Carnot engine and for the inverted

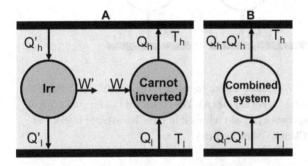

Figure 3.6   Demonstration of Carnot theorem. We assume that a heat engine *Irr* exists that is
more efficient than the Carnot engine, and combine it with an inverted Carnot engine (A). The
combined system only extracts heat from the reservoir at T_l and rejects heat into the reservoir
at $T_h > T_l$, which violates Clausius' statement (B).

arnot engine. Since $\eta_{Irr} > \eta_{Carnot}$,

$$Q_h > Q'_h.$$

y applying the first law to the combined system, it is easy to see that $(Q_h - Q'_h) =$
$2_l - Q'_l)$. Moreover, since $(Q_h - Q'_h) > 0$, we conclude that the combined system
ily extracts heat $(Q_l - Q'_l)$ from the reservoir at low temperature (T_l), and rejects
:at $(Q_h - Q'_h)$ into the reservoir at $T_h > T_l$, thus violating Clausius' statement. There-
»re, an heat engine that is more efficient than a Carnot engine does not exist. The
llowing is a corollary of Carnot's theorem, whose demonstration is left as an exer-
se to the reader.

Corollary: All reversible heat engines operating between two heat reservoirs
have the same efficiency

---

6 ENTROPY CHANGE IN REVERSIBLE PROCESSES

nce the efficiency of a Carnot engine does depend on the working substance, for
ιy Carnot cycle it is true that

$$\frac{Q_h}{Q_l} = \frac{T_h}{T_l},$$

, equivalently

$$\frac{Q_h}{T_h} + \frac{-Q_l}{T_l} = 0. \tag{3.15}$$

ais result is of critical importance because apart from determining the absolute
mperature scale, it also allows establishing a mathematical expression for entropy
ιange in reversible processes. Clausius was the first to recognise the existence of
is novel state function, which is one of the most important thermodynamic proper-
:s. As we will see, the most popular statement of the second law of thermodynam-
s determines the direction of entropy change in isolated systems. Here, we start by
·neralising (3.15) for a reversible continuous cycle L (Figure 3.7 A).

In one Carnot cycle there is only heat in the isotherms. Therefore, (3.15) can be
written as

$$\sum_{cycle} \frac{Q^{rev}}{T} = 0,$$

1ere we use the superscript *rev* to emphasise the fact that energy is transferred
versibly. We now approximate the continuous cycle L by a series of n Carnot cycles
igure 3.7 B). If the Carnot cycles are executed in the same direction (as indicated by
e arrows within each cycle), the work contributions of the adiabaths that are shared
· each pair of cycles cancels out. Therefore, for n Carnot cycles it is possible to
·ite

$$\sum_{i=1}^{n} \left(\sum_{cycle(i)} \frac{Q_i^{rev}}{T_i} \right) = 0.$$

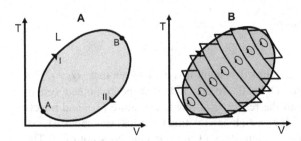

Figure 3.7   A continuous reversible cycle L represented in the (V, T) state plane showing two paths I and II (A), and the same cycle covered by Carnot cycles (B), which are all executed in the same direction.

For an infinitesimal amount of energy transferred as heat, $đQ_i$, the equation above becomes exact when $n \to \infty$. In this case:

$$\oint_L \frac{đQ^{rev}}{T} = 0 \qquad (3.16)$$

By considering paths (I) and (II) (Figure 3.7 A), one can rewrite (3.16) as

$$\int_{I_{AB}} \frac{đQ^{rev}}{T} + \int_{II_{BA}} \frac{đQ^{rev}}{T} = 0. \qquad (3.17)$$

Since, except for its sign, $đQ^{rev}$ takes on the same values in the process that goes from A to B along path II, and in the process that goes from B to A along the same path, it is true that

$$\int_{II_{BA}} \frac{đQ^{rev}}{T} = -\int_{II_{AB}} \frac{đQ^{rev}}{T}. \qquad (3.18)$$

By substituting equation (3.18) in (3.17) it comes that

$$\int_{I_{AB}} \frac{đQ^{rev}}{T} = \int_{II_{AB}} \frac{đQ^{rev}}{T}. \qquad (3.19)$$

The property expressed by (3.19) allows the definition of a novel state function. The latter, represented by the letter S, was designated by Clausius as **entropy**. The name was inspired by the Greek word *tropi*, which means change. Clausius added the prefix *en* to connect it to energy. For a reversible process that takes the system from a certain fixed equilibrium state O, to some equilibrium state A, the entropy of equilibrium state A is given by

$$S_A = S_O + \int_O^A \frac{đQ^{rev}}{T}. \qquad (3.20)$$

The equilibrium state O is called a **reference** state, and, in principle, it may be arbitrarily defined. This implies that in the context of thermodynamics entropy is not determined in absolute terms, but rather measured relative to an arbitrary value that is assigned to some conveniently chosen reference state. The SI unit of entropy is J K^{-1}.

Now, let us consider a reversible process from equilibrium state A to the reference state O, and another reversible process from reference state O to equilibrium state B. It comes that

$$\int_A^B \frac{dQ^{rev}}{T} = \int_A^O \frac{dQ^{rev}}{T} + \int_O^B \frac{dQ^{rev}}{T}.$$

Since

$$\int_A^O \frac{dQ^{rev}}{T} = -\int_O^A \frac{dQ^{rev}}{T} = S_O - S_A,$$

and the entropy of state B is

$$S_B = S_O + \int_O^B \frac{dQ^{rev}}{T},$$

the change in entropy between equilibrium states A and B is

$$\Delta S = S_B - S_A = \int_A^B \frac{dQ^{rev}}{T} \qquad (3.21)$$

If states A and B are infinitesimally separated, the equation above becomes

$$dS = \frac{dQ^{rev}}{T} \qquad (3.22)$$

For an adiabatic process $dS = 0$. For this reason, such a process is also termed **isentropic**.

Think about it...

Based on equation (3.21) can you explain why the Carnot engine has an efficiency lower than 1?

Answer

Because an efficiency equal to one would require the temperature of the heat sink to be $T = 0$ K, and, according to (3.21), no energy can be exchanged as heat with a reservoir at $T = 0$ K.

3.7   THE FUNDAMENTAL CONSTRAINT

Now that we have defined entropy for reversible processes, let us go back to the first law of thermodynamics for a fluid system with a fixed number of particles, namely,

$$dU = dQ - PdV,$$

where $-PdV$, stands for reversible work. If energy is transferred as heat reversibly, the equation above becomes

$$dU = TdS - PdV \qquad (3.23)$$

Equation (3.23) is known as the **fundamental constraint** or **fundamental equation of thermodynamics**. The variables T and S, and P and V form **conjugate pairs**. If one considers a set of generalised forces, the fundamental constraint can be written as

$$dU = TdS + \sum_i X_i dx_i, \qquad (3.24)$$

with the dot product $\sum_i \vec{X}_i \cdot d\vec{x}_i$ in the case of vectorial quantities. In a conjugate pair Xdx, one of the variables, X, is intensive and the other, x, is extensive. Note that Xdx has dimensions of energy.

As we discussed in Chapter 2, the equilibrium state of an ideal gas (or any fluid system in general) where N is fixed, stays specified by a pair of independent state variables, e.g., (T,V) or (T,P). Equation (3.23) shows that U changes whenever, S and V change. This means that $U = U(S,V)$. The difference between the (S,V) pair and other pairs of state variables, is that S and V are the *natural variables* of the internal energy. We will address this concept in the next chapter.

Given that

$$U = U(S,V), \qquad (3.25)$$

then,

$$dU = \left(\frac{\partial U}{\partial S}\right)_V dS + \left(\frac{\partial U}{\partial V}\right)_S dV. \qquad (3.26)$$

Comparing (3.26) with (3.23) it comes that

$$T = \left(\frac{\partial U}{\partial S}\right)_V \text{ and } P = -\left(\frac{\partial U}{\partial V}\right)_S.$$

The equations above provide **thermodynamic definitions** of temperature and pressure, respectively. They are **equations of state for the internal energy** because they establish a connection between the state functions, namely, $T = T(S,V)$ and $P = P(V,S)$.

Think about it...
Which of the systems in Figure 3.8 exhibits absolute negative temperature?

Answer

By using the thermodynamic definition of temperature, only system A exhibits absolute negative temperature, because it is the one where the curve representing the dependence of U on S has a negative slope.

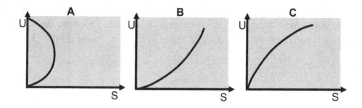

Figure 3.8 Dependence of U on S for three hypothetical thermodynamic systems.

It follows from the reciprocal rule (2.23) that

$$\frac{1}{T} = \left(\frac{\partial S}{\partial U}\right)_V,$$

and therefore,

$$S = S(U,V) \tag{3.27}$$

in a system where N is fixed. U and V are the natural variables of S. The differential of $S = S(U,V)$ is

$$dS = \left(\frac{\partial S}{\partial U}\right)_V dU + \left(\frac{\partial S}{\partial V}\right)_U dV. \tag{3.28}$$

To evaluate the second partial derivative we consider equation (3.23) with U constant. In that case

$$\left(\frac{\partial S}{\partial V}\right)_U = \frac{P}{T},$$

and,

$$dS = \frac{1}{T}dU + \frac{P}{T}dV. \tag{3.29}$$

Think about it...
Consider the Carnot cycle. How does entropy change in the isothermal expansion? And in the isothermal compression? Can you draw the Carnot cycle in the (S,T) state plane?

Answer

In both cases $dU = 0$. Using (3.29) and the ideal gas equation

$$dS = \frac{Nk_B}{V}dV.$$

Integrating between an initial equilibrium state i and a final equilibrium state f

$$\Delta S = S_f - S_i = Nk_B \ln\left(\frac{V_f}{V_i}\right).$$

Thus $\Delta S > 0$ for the isothermal expansion $(V_f > V_i)$, and $\Delta S < 0$ for the isothermal compression $(V_f < V_i)$. It is easy to see that $\Delta S_{cycle} = \Delta S_{A \to B} + \Delta S_{C \to D} = 0$, a result that could have been anticipated because S is a state function. Based on this analysis one can draw the Carnot cycle in the (T,S) state plane (Figure 3.9).

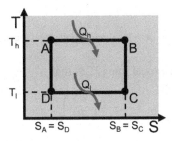

Figure 3.9   The Carnot cycle in the (S,T) state plane. The entropy increases in the isotherm at T_h, and decreases in the isotherm at T_l. From the definition of entropy change for a reversible process (3.21), the entropy is constant in the adiabaths because $Q = 0$.

Think about it...
Although V and U are the natural variables of S for a system where N is fixed, one can always express S in terms of other pairs of independent variables. As a matter of fact, two rather important thermodynamic relations can be obtained by considering $S = S(T,V)$ and $S = S(T,P)$, namely,

$$\left(\frac{\partial S}{\partial T}\right)_V = \frac{C_V}{T}, \tag{3.30}$$

and

$$\left(\frac{\partial S}{\partial T}\right)_P = \frac{C_P}{T}. \tag{3.31}$$

Can you think of a way to derive (3.30)?

Answer

If $S = S(T, V)$, then

$$dS = \left(\frac{\partial S}{\partial T}\right)_V dT + \left(\frac{\partial S}{\partial V}\right)_T dV.$$

Replacing dS as given by the equation above in equation (3.29) one obtains the following equation:

$$\frac{dU}{T} = \left(\frac{\partial S}{\partial T}\right)_V dT + \left[\left(\frac{\partial S}{\partial V}\right)_T - \frac{P}{T}\right]dV.$$

Thus, if V is constant,

$$\frac{1}{T}\left(\frac{\partial U}{\partial T}\right)_V = \left(\frac{\partial S}{\partial T}\right)_V,$$

with $C_V = \left(\frac{\partial U}{\partial T}\right)_V$.

3.8   CLAUSIUS THEOREM

According to the Carnot theorem, the efficiency of an irreversible heat engine operating between two heat reservoirs T_h and T_l is lower than the efficiency of a Carnot engine:

$$1 - \frac{Q_l}{Q_h} < 1 - \frac{T_l}{T_h}.$$

Equivalently,

$$\frac{Q_h}{T_h} + \frac{-Q_l}{T_l} < 0. \tag{3.32}$$

The Clausius theorem (or Clausius inequality) generalises the Carnot theorem for continuous irreversible cycles. Thus, equation (3.32) is a particular realisation of Clausius inequality, which we will derive in the present section. The Clausius inequality will lead to a third formulation of the second law of thermodynamics, which establishes the direction of entropy change in isolated systems.

Consider a generic thermodynamic system S that undergoes a continuous irreversible cycle *Irr* (Figure 3.10 A). The system exchanges energy Q_1, Q_2, Q_3, Q_4, ... Q_n with n heat reservoirs at temperatures T_1, T_2, T_3, T_4, ...T_n, respectively. Note that in the particular context of this demonstration we will be assuming that Q_i represents a signed number. Likewise, Q_i is positive if energy is transferred to the system, and negative if energy is transferred from the system.

Let us further consider n Carnot engines, C_1, C_2, C_3, C_4, ..., C_n. Each Carnot engine operates between a heat reservoir at temperature T_0, and another heat reservoir at temperature T_i. The size of each Carnot cycle is such that at each point of the continuous cycle *Irr*, energy Q_i is transferred from the i-th heat reservoir T_i to the system (Figure 3.10 B).

A   **B**

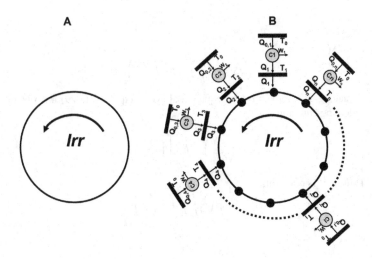

Figure 3.10   A continuous irreversible cycle *Irr* (A) is combined with *n* Carnot cycles (B). In (B) the arrows represent a *particular* realisation of energy transfer, which corresponds to positive Q_is. However, we are assuming that along the *Irr* cycle energy transfer can occur in both directions.

Think about it...

Why is the continuous cycle *Irr* in Figure 3.10 not represented in a state plane as the continuous cycle *L* of Figure 3.7A?

Answer

Because an irreversible cycle leads the system from an initial equilibrium state to a final equilibrium state through intermediate states which are non-equilibrium states. The latter cannot be represented in the state plane.

For each Carnot cycle C_i it its true that

$$\frac{Q_{0,i}}{T_0} = \frac{Q_i}{T_i},$$

or, equivalently,

$$Q_{0,i} = T_0 \frac{Q_i}{T_i}.$$

Now consider a combined cycle that comprises one complete cycle *Irr* and one complete cycle of each Carnot engine C_i. For the combined cycle, the net heat extraction at each one of the heat reservoirs T_i will be zero, because the energy that is transferred from the heat reservoir at T_i to the system, is exactly the same as that transferred to the same heat reservoir by each Carnot cycle C_i. However, the total energy transferred to the set of n Carnot cycles by the reservoirs at T_0 is:

$$Q_0 = T_0 \sum_{i=1}^{n} \frac{Q_i}{T_i}.$$

Since both the system S and the set of n Carnot cycles return to their initial state upon completion of one combined cycle, the only result of the latter is to convert Q_0 into work

$$W = W_{irr} + \sum_{i=1}^{n} W_i,$$

which would violate Kelvins' statement of the second law if Q_0 is positive. We must therefore require that Q_0 must be negative. In this case

$$\sum_{i=1}^{n} \frac{Q_i}{T_i} < 0, \tag{3.33}$$

because T_0 and T_i are measured in the absolute temperature scale (i.e. they are positive quantities). Equation (3.33) is the generalisation of (3.32) for a set of n heat transfers between the system and heat reservoirs. To derive it we considered finite energy transfers Q_i with the heat reservoirs at temperature T_i. If, instead, one considers infinitesimal energy changes dQ_i, with a continuous distribution of reservoirs such that $n \to \infty$, the equation above becomes

$$\oint \frac{dQ}{T} < 0 \tag{3.34}$$

Equations (3.16) and (3.34) are summarised in Clausius (heat) theorem.

Clausius (heat) theorem: For a thermodynamic system that exchanges heat dQ with a reservoir at temperature T, and undergoes a cyclic process

$$\oint \frac{dQ}{T} \leq 0, \tag{3.35}$$

with the equality being valid for a reversible cycle

Think about it...
What is the meaning of T in (3.35)?

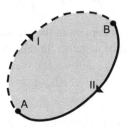

Figure 3.11   A continuous cycle showing an irreversible path (I) and a reversible one (II).

We will now consider a cycle that combines an irreversible path (I) with a reversible one (II) (Figure 3.11) (note that the cycle as a whole is considered irreversible and cannot be represented in a state plane). By applying Clausius theorem we can write

$$\int_{I_{AB}} \frac{dQ}{T} + \int_{II_{BA}} \frac{dQ^{rev}}{T} \leq 0,$$

which is equivalent to

$$\int_{I_{AB}} \frac{dQ}{T} \leq \int_{II_{AB}} \frac{dQ^{rev}}{T}.$$

Using (3.21):

$$\int_{I_{AB}} \frac{dQ}{T} \leq S_B - S_A.$$

In general, for any type of thermodynamic process between the initial state A, and the final state B

$$\int_A^B \frac{dQ}{T} \leq S_B - S_A,$$

with the equality holding for reversible processes.

When the two states are separated by an infinitesimal amount

$$\frac{dQ}{T} \leq dS.$$

The following equation summarises what we have learnt regarding entropy change:

$$dS = \frac{dQ^{rev}}{T} > \frac{dQ}{T} \qquad (3.36)$$

Equation (3.36) shows that if the system under consideration is thermally isolated $(dQ = 0)$

$$dS \geq 0. \qquad (3.37)$$

The inequality expressed by (3.37), which shows that the entropy of an isolated system cannot decrease, has an important corollary, which is the third formulation of the second law of thermodynamics:

Maximum entropy principle: In approaching thermodynamic equilibrium, the entropy of an isolated system must tend to a maximum, and the final equilibrium state is the one for which the entropy is greatest

---

Thus, **the equilibrium state of an isolated system** is such that

$$dS = 0, \qquad (3.38)$$

and

$$S = S_{max}. \qquad (3.39)$$

3.9   ANALYSIS OF IRREVERSIBLE PROCESSES

In this section we analyse in detail some irreversible processes.

3.9.1   JOULE EXPANSION

We start with the **Joule expansion**, which is also called **free expansion**. In this process an amount of ideal gas is kept in one side of a thermally isolated container via a small partition. The pressure and volume of the gas are respectively P_i and V_i. The other side of the container is under vacuum. The partition between the two parts of the container is then removed and the gas fills the whole container, ending up at a final equilibrium pressure P_f and $V_f = 2V_i$ (Figure 3.12). What is the entropy change associated with this process?

This process is clearly irreversible due to the large pressure difference that exists between both sides of the container. The equilibrium states are only the initial and final sates. Since entropy is a state function, entropy change only depends on the initial and final states of the system. Therefore, for the purpose of evaluating ΔS one can consider a reversible path taking the system from the initial to the final state for which

$$\Delta S = \int_i^f \frac{dQ^{rev}}{T}.$$

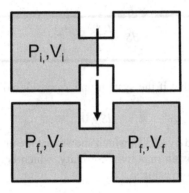

Figure 3.12 Joule expansion.

Since the whole container is thermally isolated the process is adiabatic $(Q = 0)$. Moreover, since the gas expands freely (because there is no external force opposing the gas movement), the work associated with this process is also null $(W = 0)$. The number of particles is kept constant. According to the first law of thermodynamics $\Delta U = 0$. Therefore, the reversible process that shares with the irreversible process the same initial and final states, will be an isothermal expansion since for the ideal gas $\Delta U = 0$ implies that $\Delta T = 0$. Consequently $đQ^{rev} = PdV$, and thus

$$\Delta S = \int_{V_i}^{V_f} \frac{P}{T} dV$$
$$= Nk_B \int_{V_i}^{2V_i} \frac{dV}{V}$$
$$= Nk_B \ln 2$$

Since no process occurs in the surroundings

$$\Delta S_{surroundings} = 0.$$

Thus

$$\Delta S_{universe} = \Delta S + \Delta S_{surroundings}$$
$$= Nk_B \ln 2 > 0,$$

in accordance with the second law.

3.9.2 SYSTEMS IN THERMAL CONTACT

We will now consider a particular case of two systems in thermal contact in which one of the systems is a heat reservoir R at temperature T_R, and the other is a system with heat capacity C, and temperature T_S. How to evaluate ΔS_{total}?

We start by evaluating ΔS. The macroscopic temperature difference $\Delta T = (T_R - T_S)$ leads to an irreversible energy transfer process. Since the entropy is a state function we will – as we did in the Joule expansion – use a reversible path between the initial and final states to evaluate ΔS. In this case

$$\Delta S = \int_i^f \frac{dQ^{rev}}{T},$$

with $dQ^{rev} = CdT$. Since the system is in thermal contact with a heat reservoir, the system's final temperature, will be the reservoir's temperature T_R. Thus,

$$\Delta S = C \int_{T_S}^{T_R} \frac{dT}{T} = C \ln\left(\frac{T_R}{T_S}\right).$$

To compute the entropy change of the reservoir, we need to recall that by definition of reservoir its temperature is constant. Thus,

$$\Delta S_{reservoir} = \frac{1}{T_R} \int_i^f dQ^{rev} = \frac{Q^{rev}}{T_R}$$

We also need to realise that in both the reversible and irreversible processes the amount of energy transferred to (or from) the system has the same magnitude (with opposite sign) of that transferred from (or to) the reservoir. In particular, using $Q = C\Delta T$, and taking into account that the system's final temperature is that of the reservoir, the system's temperature change between the final and initial states is $\Delta T = (T_R - T_S)$. Therefore, in the expression above, $Q^{rev} = -C(T_R - T_S)$ is the energy transferred from (to) the reservoir. Thus,

$$\Delta S_{reservoir} = \frac{C(T_S - T_R)}{T_R}.$$

Finally,

$$\Delta S_{universe} = \Delta S + \Delta S_{reservoir}$$

$$= C \ln\left(\frac{T_R}{T_S}\right) + \frac{C(T_S - T_R)}{T_R}$$

$$= C\left[\frac{T_S}{T_R} - \ln\left(\frac{T_S}{T_R}\right) - 1\right].$$

Since $C > 0$, it is easy to see by plotting the function $(x - \ln x - 1)$ that $\Delta S_{universe} > 0$.

3.9.3 MIXING TWO IDEAL GASES

Let us consider two different ideal gases, 1 and 2, at the same temperature T and pressure P, and occupying volumes $(1-x)V$ and Vx as represented in Figure 3.13 A. The partition that separates the two gases is removed, and each gas expands ending up occupying the whole volume V. What is the entropy change associated with this process? Since the process occurs at fixed temperature, the internal energy does not

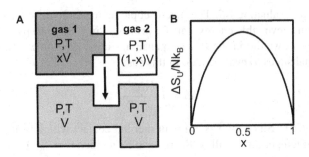

Figure 3.13   Mixing two ideal gases (A), and the corresponding entropy change (B).

change. Therefore, we can consider an isothermal reversible expansion for each gas to evaluate the corresponding entropy change. For gas 1 we get

$$\Delta S_1 = x k_B \int_{xV}^{V} \frac{1}{V_1} dV_1,$$

and for gas 2,

$$\Delta S_2 = (1-x) k_B \int_{(1-x)V}^{V} \frac{1}{V_2} dV_2.$$

Since there is nothing occurring elsewhere, the entropy change of the universe is

$$\Delta S_{universe} = \Delta S_1 + \Delta S_2 = -N k_B \left(x \ln x + (1-x) \ln(1-x) \right).$$

A plot of $\Delta S_{universe}$ is represented in Figure 3.13 B, where it is possible to see that the maximum entropy change corresponds to $x = 0.5$.

Think about it...

If the two gases are identical what is $\Delta S_{universe}$?

Answer

In that case there is no thermodynamic process because removing the partition has no observable macroscopic consequences. Therefore, $\Delta S_{universe} = 0$.

3.10   THE STATISTICAL MEANING OF ENTROPY

In thermodynamics the equilibrium state of a system, the so-called **equilibrium macrostate**, stays completely specified by a subset of its macroscopic properties. Indeed, in the case of a simple fluid with a fixed number of particles, any pair of variables (T,V), (T,P) and so forth, is enough to completely determine the system's macrostate (Figure 3.14 A). From a microscopic point of view, however, the fluid is an ensemble of dynamic N point particles, and each particle i has a position vector, $\vec{r} = (x_i, y_i, z_i)$ and a linear momentum vector $\vec{p} = (mv_{ix}, mv_{iy}, mv_{iz})$. Therefore, the ensemble of N particles stays completely characterised by a point in the $6N$ dimensional space called the **phase space**. Each point in the phase space represents a **microstate** of the system.

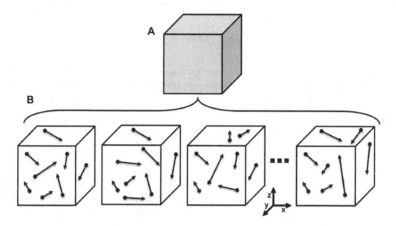

Figure 3.14   A macrostate of a thermodynamic system (A), and several microstates compatible with the macrostate (B).

In a famous paper dated from 1872 Ludwig Boltzmann showed that in thermodynamic equilibrium the entropy is given by

$$S = k_B \ln W, \tag{3.40}$$

where W, the first letter of the german word *wahrscheinlichkeit* (probability), represents the number of microstates compatible with a given equilibrium macrostate with fixed energy $U = E$, volume V, and number of particles N (Figure 3.14 B). Notice that while W is generally called the probability of a given macrostate, in fact, it is only proportional to the probability in the usual statistical sense. This famous equation, despite reflecting Boltzmann's ideas, was actually written by Max Planck (1858–1974) in 1900, and is craved in Boltzmann's tomb.

The second law of thermodynamics states that in an isolated system entropy tends to a maximum, and is maximum in equilibrium. Boltzmann's equation allows to understand why this happens: An isolated system tends toward an equilibrium macrostate with maximum entropy, because then the number of microstates is the largest, which makes this state the statistically most likely.

Boltzmann equation also allows us to understand the reason why entropy increases in Joule expansion, a process that occurs at constant energy (of both the system and the thermodynamic universe). In the Joule expansion, nothing occurs in the surroundings. Thus we need to focus our analysis only on the system. In the final equilibrium state, each particle of the gas has access to a volume which is twice as larger as that of the initial state, $V_f = 2V_i$. Therefore, the number of microscopic configurations (i.e. spatial arrangements of the particles) within the available volume is also larger. In other words, the number of system's microstates compatible with the equilibrium macrostate characterised by some V is also larger. The final state is thus the one that maximises the number of microstates of the thermodynamic universe. This reasoning allows one to understand the dependence on V in the equation

$$dS = \frac{P}{T}dV + \frac{dU}{T}.$$

To understand the dependence on U, we need to consider a non-isolated system, but take into account that the thermodynamic universe will itself be isolated. Let us consider the case of two systems in thermal contact. More precisely, let an ideal gas enclosed by some container, be placed into thermal contact with a heat reservoir. In this case since the volume is kept fixed there is no energy transfer as work, and the only way to change the internal energy of both the system and reservoir is by transferring energy as heat. Again, the final state should be the one that maximises the number of microstates of the thermodynamic universe. At this point it is useful to recall that the energy of a system in contact with a thermal bath is not fixed (only the temperature is fixed) fluctuating around an equilibrium value. Indeed, the energy is a random variable to which corresponds a probability distribution which depends on the temperature T of the heat reservoir. In Figure 3.15 we show the probability distribution function of the energy of an ideal gas (which is only kinetic energy) at three different temperatures (see problem 1.15). We see that the distribution becomes broader as T increases. This means that at higher T the system has access to a larger number of microstates. This information allows one to understand why the entropy of a system increases when it is placed with a heat reservoir with $T_R > T_S$. In the opposite case, when $T_R < T_S$, energy will be transferred as heat from the system to the reservoir, and the entropy of the system decreases. However, the entropy of the thermodynamic universe, and, in particular, the number of microstates of the reservoir, must necessarily increase according to the second law.

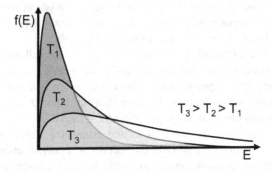

Figure 3.15   Distribution of energy of an ideal gas in thermal contact with three reservoirs.

3.11   ENTROPY AND DISORDER

It is often said that entropy is a measure of **disorder**. But what does disorder actually mean? It appears that Hermann von Helmholtz (1821–1894) was the first to establish the association between entropy and disorder in a paper on the kinetic theory of gases from 1882 where, at some stage, it reads: *Unordered [ungeordnete] motion, in contrast, would be such that the motion of each individual particle needs to have no similarity to that of its neighbours. We have ample ground to believe that heat-motion is of the latter kind, and one may in this sense characterise the magnitude of the entropy as the measure of the disorder [unordnung].* In this sentence, Helmholtz associates the unordered motion of the gas to a lack of structural order, and considers that energy transferred as heat to the gas (which leads to an entropy increase) should result into this kind of unordered motion.

Contrary to a gas, a solid exhibits long-range order, which results in a regular arrangement of particles which repeats itself periodically over the entire crystal. Considering the Boltzmann's definition of entropy, the number of microscopic configurations compatible with a certain volume, energy and number of particles of the system will be much higher for the gas state than for the solid state, where the particles are constrained to the vertices of a lattice, vibrating around their fixed positions. In this sense, it seems reasonable to consider entropy a measure of **structural disorder**, with the more structurally disordered state being the one with higher entropy. However, this association may not always be correct. An illustrative example from the macroscopic world helps to understand the reason it may fail for some physical systems. Consider packing a rigid (i.e. fixed volume V) suitcase, when the number of items is such that their combined volume is similar to that of the suitcase (in thermodynamic terms this would be a high-density system). In this scenario, one cannot simply randomly place the items inside the suitcase (as we probably would if the available volume was larger), but one has to place them in an orderly manner to be able to close it. Therefore, in this case, there are more structurally organised arrangements compatible with the suitcase's volume than disorganised ones. If one considers that an arrangement of the suitcase corresponds to a "microstate" compatible with volume V, then the number of structurally organised microstates is larger than the number of corresponding disordered microstates. Therefore, the entropy of the structurally ordered state is higher than that of the structurally disordered one. What this example suggests is that there may be physical systems at high density, where packing considerations dominate, and in which the more ordered phase possesses a higher entropy (i.e. a higher number of microstates) than a spatially disordered phase at the same temperature and density. This was originally observed in molecular simulations of a fluid modelled as a system of hard spheres undergoing a freezing transition, which is known to provide a good description of certain classes of colloidal systems. For these systems, the more efficient packing of the ordered phase leads to both a greater free volume and local mobility, which overcomes the effect of increased positional ordering in determining the entropy at high density.

While the view of entropy as a measure of structural disorder is the most common, sometimes the word disorder interpreted as **lack of information** is also employed in association to entropy. The justification in this case being that the greater the number microstates compatible with a given equilibrium macrostate, the less information is available about the precise microstate, i.e, the higher the uncertainty associated with the occurrence of a given microstate.

Entropy is a precise, measurable quantity, and while the associations described above may be tempting, it is important to use them correctly (i.e. for the physical systems where they actually apply) so that they help us understand instead of hindering the correct meaning of entropy.

3.12 LEARNING OUTCOMES

At the end of this chapter the reader is expected to:

1. Be able to quantitatively analyse the Carnot cycle and other cyclic processes used to operate heat engines.
2. Understand the importance of the Carnot engine for the formulation of the second law of thermodynamics.
3. Be able to prove the logical equivalence between Kelvin's and Clausius' statements of the second law.
4. Be able to prove Carnot's theorem.
5. Understand that the relation $\frac{Q_h}{Q_l} = \frac{T_h}{T_l}$ determines the absolute (or thermodynamic) temperature scale.
6. Establish the definition of entropy change for reversible processes by generalising the relation $\frac{Q_h}{Q_l} = \frac{T_h}{T_l}$ to continuous reversible cycles.
7. Appreciate the importance of the fundamental equation of thermodynamics, namely, that it establishes the equations of state $T = T(S,V)$ and $P = P(V,S)$.
8. Understand that Clausius theorem generalises Carnot's theorem for continuous irreversible cycles and be able to demonstrate it.
9. Appreciate the importance of Clausius theorem and use it to establish the maximum entropy principle.
10. Know how to compute entropy change for different types of irreversible processes.
11. Know the statistical significance of entropy, and understand the connection between entropy and disorder.

3.13 WORKED PROBLEMS

PROBLEM 3.1
The Stirling cycle that uses the ideal gas as working substance consists of the following processes: 1) isothermal compression from state (P_1, V_1) to state (P_2, V_2), 2) heating at constant volume from (P_2, V_2) to (P_3, V_2), 3) isothermal expansion from (P_3, V_2) to (P_4, V_1), and 4) cooling at constant volume back to (P_1, V_1).
a) Draw the Stirling cycle in the (P,V) state plane.
b) Determine the sign of Q, W, and ΔS after one cycle.

c) What is the efficiency of a engine operating on the basis of this cycle?

Solution

a) The Stirling cycle is represented in Figure 3.16.
b) In the isotherms ($dU = 0$):

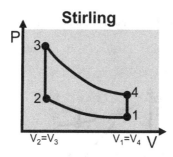

Figure 3.16   The Stirling cycle, where the processes $1 \to 2$ and $3 \to 4$ are isotherms at T_l and $T_h > T_l$, respectively.

$$Q_{1\to2} = Nk_B T_l \ln\left(\frac{V_2}{V_1}\right) < 0,$$

and

$$Q_{3\to4} = Nk_B T_h \ln\left(\frac{V_4}{V_3}\right) = Nk_B T_h \ln\left(\frac{V_1}{V_2}\right) > 0.$$

In the isochors ($dQ = dU$). Thus

$$Q_{2\to3} = C_V (T_h - T_l) = -Q_{4\to1} = C_V (T_h - T_l).$$

Therefore,

$$Q_{cycle} = Nk_B T_l \ln\left(\frac{V_2}{V_1}\right) + Nk_B T_h \ln\left(\frac{V_1}{V_2}\right)$$
$$= Nk_B \left(\frac{V_1}{V_2}\right)(T_h - T_l) > 0.$$

Since $\Delta U_{cycle} = 0$,

$$W_{cycle} = -Q_{cycle} = Nk_B \left(\frac{V_2}{V_1}\right)(T_l - T_h) < 0.$$

Since S is a state function, $\Delta S_{cycle} = 0$.
   c) Since this is a cycle operating between two heat reservoirs, the efficiency of an engine that operates this cycle must be the same as that of the Carnot engine:

$$\eta = 1 - \frac{T_l}{T_h}.$$

To confirm it, let us use the definition of efficiency of a heat engine, i.e., the ratio between the and work produced after one cycle ($W = -W_{cycle}$), and heat Q_h extracted from the heat reservoir at T_h,

$$\eta = \frac{W}{Q_h} = \frac{Nk_B\left(\frac{V_2}{V_1}\right)(T_h - T_l)}{Nk_B T_h \ln\left(\frac{V_1}{V_2}\right)} = 1 - \frac{T_l}{T_h}.$$

PROBLEM 3.2

Prove that two reversible adiabaths cannot intersect.

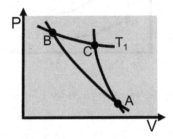

Figure 3.17   Two hypothetical adiabaths that intersect at point A.

Solution

Assuming that two adiabaths intersect at some point A on the (P,V) state plane, it is possible to draw an isotherm at $T = T_1$ that intersects the adiabaths at points B and C as shown in Figure 3.17. The corresponding cycle extracts heat from the reservoir at T_1 without producing any other effect, in clear violation of the second law of thermodynamics. Therefore, two reversible abiabaths cannot intersect.

PROBLEM 3.3

Consider an arbitrary heat engine A that operates between systems 1 and 2, working as heat reservoirs, which have identical heat capacity C independent of temperature. Let T_1 and $T_2 > T_1$ be the initial temperatures of the reservoirs. The heat engine will work until the two heat reservoirs attain the same final temperature T_3. Assume that the bodies do not contract nor expand upon changing their temperatures. Show that $T_3 > \sqrt{T_1 T_2}$ and determine the maximum work produced by the engine.

Solution

In this problem the systems acting as reservoirs have a finite heat capacity (i.e. they are not ideal reservoirs). This means that their temperature will change as a result of energy being extracted from 2 and rejected into 1 by the heat engine. If one considers the isolated system formed by the two reservoirs and the heat engine, the second law of thermodynamics implies that after one cycle

$$\Delta S = \Delta S_1 + \Delta S_2 \geq 0,$$

where we have used the fact that after one cycle $\Delta S_A = 0$, and the equality refers to a reversible process.

To evaluate the entropy change of the reservoirs we consider a reversible path for which

$$dS = \frac{dQ^{rev}}{T},$$

with $dQ^{rev} = CdT$. Thus

$$dS = dS_1 + dS_2 \geq 0.$$

Taking into account that both reservoirs equilibrate at temperature T_3, we can integrate the equation above between T_1 and T_3 in the case of reservoir 1, and between T_2 and T_3 in the case of reservoir 2 to get:

$$\Delta S = C \int_{T_1}^{T_3} \frac{dT}{T} + C \int_{T_2}^{T_3} \frac{dT}{T} \geq 0$$
$$= C \ln\left(\frac{T_3}{T_1}\right) + C \ln\left(\frac{T_3}{T_2}\right) \geq 0$$
$$= \ln\left(\frac{T_3^2}{T_1 T_2}\right) \geq 0,$$

where we have used the fact that C is constant. The last inequality implies that

$$\frac{T_3^2}{T_1 T_2} \geq 1,$$

and, therefore, $T_3 \geq \sqrt{T_1 T_2}$.

To compute the maximum work produced by engine A, the cycle must be reversible. In this case $\Delta S = 0$ and $T_3 = \sqrt{T_1 T_2}$. In a reversible cycle of A, $\Delta U_A = 0$, and according to the first law:

$$Q_h - Q_l = W_A^{rev}.$$

Let Q_2 be the energy that is extracted from the reservoir at T_2 and Q_1 the energy that is rejected into the reservoir at T_1.

Since $dQ = CdT$,

$$Q_2 = \int_{T_2}^{T_3} CdT = C(T_3 - T_2) = C(\sqrt{T_1 T_2} - T_2) < 0,$$

and

$$Q_1 = \int_{T_1}^{T_3} CdT = C(T_3 - T_1) = C(\sqrt{T_1 T_2} - T_1) > 0.$$

Thus, the energy transferred to the system from the reservoir at T_2 is

$$Q_h \equiv -Q_2,$$

and the energy rejected by the system to the reservoir at T_1 is

$$Q_l \equiv Q_1.$$

Therefore

$$W_A^{rev} = C(T_1 + T_2 - 2\sqrt{T_1 T_2}).$$

PROBLEM 3.4
Consider equation 3.29 and show that the entropy of a gas composed of N particles can be written as

$$S = C_V \ln T + Nk_B \ln V + c, \qquad (3.41)$$

with c being a constant.

Solution

Considering that for the ideal gas $dU = C_V dT$ and that $P/T = Nk_B/V$, equation (3.29) becomes

$$dS = \frac{C_V}{T} dT + \frac{Nk_B}{V} dV. \qquad (3.42)$$

Integrating equation (3.42) at constant temperature from state $S(T_O, V_O)$ to state $S(T_O, V)$ one gets

$$S(T_O, V) = S(T_O, V_O) + \int_{V_O}^{V} \left(\frac{\partial S}{\partial V} \right)_T dV$$
$$= S(T_O, V_O) + Nk_B \ln \left(\frac{V}{V_O} \right).$$

Subsequently, integrating (3.42) at constant volume from state $S(T, V_O)$ to $S(T, V)$ one gets

$$S(T, V) = S(T, V_O) + \int_{T_O}^{T} \left(\frac{\partial S}{\partial T} \right)_V dT$$
$$= S(T, V_O) + C_V \ln \left(\frac{T}{T_O} \right)$$
$$= S(T_O, V_O) + Nk_B \ln \left(\frac{V}{V_O} \right) + C_V \ln \left(\frac{T}{T_O} \right)$$
$$= Nk_B \ln V + C_V \ln T + S(T_O, V_O) - NK_B \ln V_O - NK_B \ln T_O.$$

Finally one can write

$$S(T, V) = C_V \ln T + Nk_B \ln V + c,$$

with $c = S(T_O, V_O) - NK_B \ln V_O - NK_B \ln T_O.$

3.14   SUGGESTED PROBLEMS

PROBLEM 3.5
What is the maximum efficiency of a heat engine operating between $T = 20°C$ and $T = 500°C$.

PROBLEM 3.6
If a real engine operating between $T = 20°C$ and $T = 500°C$ produces 120 J of work and rejects 180 J of heat into the colder reservoir, what is the efficiency of this engine?

PROBLEM 3.7
Consider the ideal gas cycle represented in Figure 2.5. Show that a heat engine operating on the basis of this cycle has an efficiency given by

$$\eta = 1 - \gamma \frac{\left(\frac{V_1}{V_2}\right) - 1}{\left(\frac{P_2}{P_1}\right) - 1},$$

where γ is the adiabatic index.

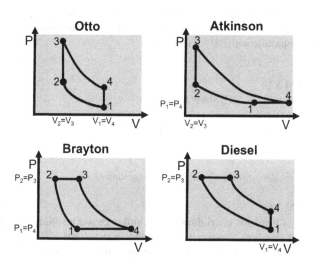

Figure 3.18   Four cycles of the ideal gas. In all the cycles the processes $1 \to 2$ and $3 \to 4$ are adiabaths.

PROBLEM 3.8
Consider the four cycles of the ideal gas represented in Figure 3.18 where the processes $1 \to 2$ and $3 \to 4$ are adiabaths. Show that:

a) The efficiency of a engine operating on the basis of the Otto cycle is

$$\eta = 1 - \left(\frac{V_1}{V_2}\right)^{(1-\gamma)}.$$

b) The efficiency of a engine operating on the basis of the Brayton cycle is

$$\eta = 1 - \left(\frac{T_1}{T_2}\right).$$

c) The efficiency of a engine operating on the basis of the Atkinson cycle is

$$\eta = 1 - \gamma \frac{(\alpha - r)}{(\alpha^\gamma - r^\gamma)}$$

with $r = V_1/V_2$ and $\alpha = \frac{V_3}{V_2}$.

d) The efficiency of a engine operating on the basis of the Diesel cycle is

$$\eta = 1 - \frac{1}{\gamma}\left(\frac{r_e^{-\gamma} - r^{-\gamma}}{r_e^{-1} - r^{-1}}\right),$$

with $r = V_1/V_2$ and $r_e = V_1/V_3$.

PROBLEM 3.9
In the Joule expansion $\Delta U = 0$ and $W = 0$. However, the entropy increases in agreement with what is expected for an irreversible process. Are these results compatible with $dU = TdS - PdV$?

PROBLEM 3.10
Starting from $S = S(T,P)$ show that

$$\left(\frac{\partial S}{\partial T}\right)_P = \frac{C_P}{T}.$$

Hint: Use the definition of C_P as given by equation (2.20).

PROBLEM 3.11
One mole of ideal gas is originally at T_0 and occupies a volume V_0. The gas expands at constant temperature, and the final volume is $2V_0$. Show that $\Delta S > 0$, and explain why the entropy increases in this process.

PROBLEM 3.12
An ideal gas is placed inside a cylinder tapped by a piston. The gas is compressed by moving the piston very slowly while the temperature is kept constant at $20°C$. During the compression the system consumes work corresponding to 730 J. Determine ΔS.

PROBLEM 3.13

An ideal gas composed of N particles is placed inside a cylinder with adiabatic walls tapped by a piston. The initial volume is V_1, and the initial temperature is T_1. Determine ΔT, ΔP, and ΔS that result from a process where the volume suddenly increases to V_2 upon removing the piston.

PROBLEM 3.14

Taking into account that the average velocity of the gas particles is

$$\langle v \rangle = \sqrt{\frac{8k_B T}{\pi m}},$$

how should we move the piston so that the equation $dS = (P/T)dV$ applies during the whole process?

PROBLEM 3.15

Let Q be the energy transferred as heat from one reservoir at temperature T_1 to another reservoir at temperature $T_2 > T_1$. The heat capacity of the reservoirs is so large, so that their temperatures do not change. Show that the entropy change associated with this process is positive.

PROBLEM 3.16

Two reservoirs have internal energy $U = CNT$, with C being constant, N the number of particles, and T the temperature. The initial temperatures of the reservoirs are T_1 and T_2. They are connected with a Carnot engine to produce work until they relax to a final temperature T_f. Determine T_f.

PROBLEM 3.17

One kg of water at $T = 0°C$ is placed into contact with a reservoir at $T = 100°C$. When the water's temperature is $T = 100°C$. What is ΔS_{H_2O}? And $\Delta S_{universe}$? How should one heat the water up to $T = 100°C$, so that $\Delta S_{universe} = 0$?

REFERENCES

1. Barón, M. (1989). With Clausius from energy to entropy. J. Chem. Education 66:1001-1004.

2. Blundell, S. J. & Blundell, K. J. (2009). Concepts in Thermal Physics Oxford University Press.

3. Cropper, W. H. (1986). Rudolf Clausius and the road to entropy. Am. J. Phys. 54: 1068.

4. Erlichson, H. (1999). Sadi Carnot, 'Founder of the second law of thermodynamics. Eur. J. Phys. 20: 183–192.

5. Fermi, E. (1956). Thermodynamics. Dover.

6. Frenkel, D. (1999). Order through disorder: Entropy-driven phase transitions. In: Garrido L. (eds) Complex Fluids. Lecture Notes in Physics 415. Springer, Berlin, Heidelberg.

7. Laird, B.B. (1999). Entropy, disorder and freezing. J. Chem. Educ. 76:1388-1390.

8. Lemmons, D. S. (2009). Mere Thermodynamics. Johns Hopkins University Press.

9. Lemmons, D. S. & Penner M. K. (2008). Sadi Carnot's contribution to the second law of thermodynamics. Am. J. Phys. 76:21-25.

10. Sharp, K.& Matschinsky, F. (2015). Translation of Ludwig Boltzmann's Paper "On the relationship between the second fundamental theorem of the mechanical theory of heat and probability calculations regarding the conditions for thermal equilibrium." Entropy 17:1971-2009.

11. Steyer, D. (2019). Entropy as disorder: History of a misconception. The Physics Teacher 57: 454-458.

12. Swendsen, R. H. (2012). An Introduction to Statistical Mechanics and Thermodynamics. Oxford University Press.

4 The Third Law

This chapter terminates the first part of this book that is dedicated to the laws of thermodynamics. We have extensively used the first law in the analysis that led us to the formulation of the second law, and, in particular, to the establishment of the maximum entropy principle. The third law, on the other hand, is completely independent of the other laws of thermodynamics. The major goal of this chapter is to introduce the two most important formulations of the third law and show that they are logically equivalent.

4.1 INTRODUCTION

The third law of thermodynamics is about the behaviour of physical systems as the temperature approaches absolute zero. We have already seen that $T = 0$ K lies on the boundary of ordinary thermodynamic parameter space. T shows up in the denominator of many equations we have written down, and if, for instance, a Carnot cycle could be made to operate between a heat source at temperature T_h and a heat sink at temperature $T_l = 0$ K, its efficiency would be $\eta_{Carnot} = 1$ (3.7) and so it would completely convert heat into work, violating Kelvin's statement of the second law.

The third law goes beyond this observation. It reflects experimental results obtained for many different substances, and cannot be deduced from the first and second laws. Firstly stated in 1906 by Walther Nernst (1864–1941) as an independent universal principle, it has known since then several formulations. There is still on-going debate about the equivalence of different formulations and the universal validity of some of them. In this sense, it does not share the status of the first and second laws.

4.2 NERNST THEOREM

Let us begin by considering the following formulation of the third law:

Nernst (heat) theorem: The entropy change of a system in any reversible isothermal process tends to zero as the temperature of the process tends to absolute zero

---

This means that, for any extensive parameter X,

$$\Delta S = S(T,X_1) - S(T,X_2) \to 0, \text{ as } T \to 0,$$

and so, assuming that $S(T,X)$ remains finite and is continuous in the limit $T \to 0$, the entropy must approach a limiting value that is independent of the other thermodynamic variables:

$$S(0,X_1) = S(0,X_2) = S_0. \tag{4.1}$$

DOI: 10.1201/9781003091929-4

Another consequence of assuming a finite limit for the entropy is that the heat capacity will tend to zero when cooling down to absolute zero. Indeed,

$$C_X = T\left(\frac{\partial S}{\partial T}\right)_X = \left(\frac{\partial S}{\partial \ln T}\right)_X \to 0, \text{ as } T \to 0, \tag{4.2}$$

because when $T \to 0$, $\ln T \to \infty$, and $S \to S_0$.

Both Nerst's theorem and (4.2) were supported by extensive and very precise thermochemical measurements performed by Nernst on many different substances at low temperatures. The heat capacity measurements inspired Albert Einstein (1879–1955) to develop his quantum model for the heat capacity of solids in 1907. Later, in 1912, Peter Debye (1884–1966) improved Einstein's model of an insulating solid to obtain a better fit to Nernst's results, with $C \propto T^3$ at very low temperatures.

For his work on this subject, Nernst received the 1920 Nobel Prize in Chemistry. However, while the heat theorem may be valid for many systems, it is unlikely to be universal. Indeed, there are quantum systems for which the ground state (i.e. the state of minimum energy) is degenerate, and the degeneracy of the ground state may depend on the external parameter X. Since the degeneracy of the ground state determines the entropy according to Boltzmann's entropy equation (3.40), for these systems, the entropy S is not independent of X at $T = 0$ K.

4.3   MAXIMUM COOLING

Before we discuss another formulation of the third law, we analyse the process of cooling a system, and how maximum cooling can be achieved.

In principle, an adiabatic process should be an efficient means to cool down a thermodynamic system as it ensures that no heat will flow into the system when the temperature of the surroundings is higher than the system's temperature. From the analysis of the Carnot cycle we know that a reversible adiabatic expansion ($dQ = 0$, $dV > 0$) is a process that lowers the temperature of the ideal gas. Does it lead to maximum cooling?

In order to answer this question let us consider that the state of a system is described by the independent variables S and V. In this case $T = T(S,V)$, and the differential change in temperature during an expansion is given by

$$dT = \frac{T}{C_V}dS + \left(\frac{\partial T}{\partial V}\right)_S dV. \tag{4.3}$$

The first term on the right hand-side is positive because T is positive, and C_V (and, in fact, any other heat capacity) is always positive for the equilibrium state to be stable (see section 5.11.1). Thus, cooling by expansion will be maximised when $dS = 0$, i.e., for an adiabatic reversible expansion. The same conclusion holds for more general systems which can do other kinds of work apart from expansion work: maximum cooling is achieved by a reversible adiabatic process.

4.4 NERNST UNATTAINABILITY PRINCIPLE

With the final remark of the previous section in mind, it is reasonable that the following formulation of the third law, also first stated by Nernst, should be called the **unattainability principle.**

Nernst (unattainability) principle: No isentropic process starting at a non-zero temperature can take the system to zero temperature

---

The unattainability principle is sometimes taken as stating that it is not possible to cool any system to absolute zero temperature in a finite number of isothermic and isentropic reversible changes, since such a finite sequence would have to end with an isentropic cooling from a non-zero temperature down to zero. However, in spite of its name, the unattainability principle does not forbid reaching zero temperature following a different equilibrium path or a non-equilibrium process.

4.5 LOGICAL EQUIVALENCE BETWEEN NERNST THEOREM AND THE UNATTAINABILITY PRINCIPLE

The logical relation between these the two formulations of the third law is readily understood by considering the two panels of Figure 4.1, where the slope of $S = S(T, X_i)$, $i = 1, 2$, must be positive, because (4.2) holds and the heat capacity is positive.

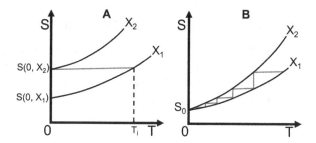

Figure 4.1   The dependence of entropy on temperature for two different values of the parameter X (e.g. V). (A) The heat theorem is violated and there is an isentropic curve connecting (T_i, X_1) to $(0, X_2)$. (B) The heat theorem is satisfied, and an infinite number of isentropic processes $S(T_i, X_1) \to S(T_f, X_2)$ combined with isothermal changes $S(T_f, X_2) \to S(T_f, X_1)$ would be necessary to cool the system down to absolute zero.

Assume first that the unattainability principle does not hold, and that there is an isentropic process connecting (T_i, X_1) to $(0, X_2)$, so that $S(T_i, X_1) = S(0, X_2)$ (Figure 4.1 A). Since the slope of $S = S(T, X_1)$ is positive, we must have $S(T_i, X_1) > S(0, X_1)$, and so $S(0, X_2) > S(0, X_1)$, in contradiction with the heat theorem formulation. Therefore, the heat theorem implies the unattainability principle, and reaching zero temperature requires an infinite number of isothermic and isentropic reversible changes, as illustrated in Figure 4.1 B.

Assume now that the heat theorem does not hold, and so the distance between the curves $S(T,X_1)$ and $S(T,X_2)$ tends to a finite entropy difference as T tends to zero. Then, as shown in Figure 4.1 A, we have $S(0,X_2) > S(0,X_1)$, and the curve emanating from $S(0,X_1)$ must intersect the isentrope emanating from $S(0,X_2)$ at some T_i. This is an isentropic process connecting (T_i,X_1) to $(0,X_2)$, which contradicts the unattainability principle.

4.6 OPEN QUESTIONS

The two statements of the third law are then formally equivalent, but some questions remain from the physical point of view. Can other cooling protocols be devised that allow zero temperature to be reached when the heat theorem formulation holds? Is the isentropic process of Figure 4.1 A physically possible for systems that violate the heat theorem formulation? In particular, the connection between attainability of zero temperature and the cooling time is not explicit in these formulations, although the infinite number of cooling steps required by the unattainability principle must involve an infinite time too.

A study by Oppenheim and Masanes published in 2017 revisits these issues, providing universal lower bounds for the attainable temperature as a function of cooling time. In doing so, the authors also showed that it is possible to violate the heat theorem without violating the unattainability principle, the latter stated as the impossibility of reaching zero temperature by any process in finite time.

4.7 THE IDEAL GAS NEAR ABSOLUTE ZERO

While no real system will ever reach $T = 0\,\mathrm{K}$, it is still possible to get close to absolute zero. Therefore, it is important to analyse the behaviour of physical systems and their models in this limit. Consider the case of an ideal gas. As shown in solved problem 3.3, the entropy of the ideal gas can be written as

$$S = C_V \ln T + Nk_B \ln V + c,$$

with $C_V = \frac{3}{2}Nk_B$ and c another constant independent of T and V. Thus, as $T \to 0$, $\ln T \to -\infty$ and $S \to -\infty$. However, C_V remains finite, in contradiction with the experimental results that lead to the third law. This means, of course, that the ideal gas model breaks down at very low temperatures, a fact that should not come as a surprise, since intermolecular interactions that become relevant at low T are not taken into account in the model. However, we shall see in Chapter 7 a more sophisticated model of a gas, developed by van der Waals, which considers intermolecular interactions still fails to predict any dependence of the heat capacity on temperature. Indeed, quantum effects that are negligible at high temperatures must be considered in order to correctly capture the low temperature behaviour of any substance.

A model in physics is only valid within the limits of its applicability and the third law highlights a fundamental limitation of classical models.

4.8 LEARNING OUTCOMES

At the end of this chapter the reader is expected to:

1. Know the Nernst heat theorem.
2. Know that the third law of thermodynamics derives from the Nernst heat theorem, being independent of the other laws of thermodynamics.
3. Understand the importance of the heat theorem in the calculation of absolute entropies.
4. Know the unattainability principle.
5. Be able to prove the logical equivalence between the heat theorem and the unattainability principle.

REFERENCES

1. Klimenko, A. Y. (2012).Teaching the thid law of thermodynamics. The Open Thermodynamics Journal 6:1-14.

2. Landsberg, P. T. (1978).Thermodynamics and Statistical Mechanics. Dover.

3. Masanes L. & Oppenheim, J. (2017). A general derivation and quantification of the third law of thermodynamics. Nat. Commun. 8:14538.

Section II

The Structure of
Thermodynamics

5 Thermodynamic Potentials

This chapter is dedicated to formal aspects of thermodynamics. It starts by physically motivating the need for special thermodynamic potentials called free energies. The Legendre transform is presented to understand the formal origin of thermodynamic potentials. The chemical potential is defined, and the fundamental equation of thermodynamics is presented for single and multicomponent systems. The chemical potential motivates the formulation of thermodynamics in the internal energy representation, and entropy representation. Massieu functions are presented as the Legendre transforms of the entropy. The conditions for thermodynamic equilibrium are discussed. The criteria for thermal and mechanical stability are established based on the analysis of local stability of equilibrium states.

5.1 INTRODUCTION

The second law of thermodynamics, expressed by the principle of maximum entropy, tells us that an isolated system tends to an equilibrium state that is characterised for having maximum entropy. We also learnt from Clausius inequality (3.36) that when the system is not thermally isolated, the entropy change associated with a thermodynamic process is such that

$$\dd Q \leq T dS,$$

with the equality applying to a reversible process.

On the other hand, the first law states that $\dd Q = dU + PdV$, and therefore

$$dU \leq T dS - P dV. \tag{5.1}$$

Equation (5.1) shows that if the entropy is constant and the volume is kept fixed during a thermodynamic process occurring in a closed system, then $dU \leq 0$. Since U is constant at equilibrium ($dU = 0$), the system will relax towards an equilibrium state characterised by a minimum value of the internal energy, U_{min}.

The following principles resume what we have learned so far:

1. **Maximum entropy principle**: The equilibrium state of an **isolated** system is the one that **maximises the entropy** ($dS \geq 0$).
2. **Minimum energy principle**: The equilibrium state of a closed system, with **constant entropy** and **fixed volume** is the one that **minimises the internal energy** ($dU \leq 0$).

The statements expressed by 1 and 2 are known as **extremum principles**, because S and U take on extremum values at equilibrium. The extremum principles are part of the reason why S and V are designated as **natural variables** of the internal energy, and U and V are the **natural variables** of the entropy. A graphical representation of the extremum principles is provided in Figure 5.1.

DOI: 10.1201/9781003091929-5   103

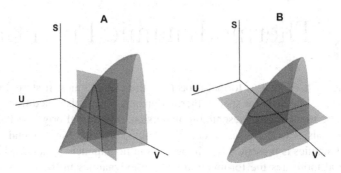

Figure 5.1 Geometrical illustration of the principle of maximum entropy (A), where the equilibrium state is the point of maximum S for constant U, and principle of minimum energy (B), where the equilibrium state is the point of minimum U for constant S.

Since an isolated system is not allowed to interact with the surroundings (i.e. it cannot be externally perturbed), it is of no interest from an experimental point of view. Furthermore, while it may be easy to keep the system's volume fixed, no practical instruments exist to control and measure entropy, which makes the extremum principle for the internal energy of no practical use either. On the other hand, in the laboratory we have the possibility to control other state functions such as the temperature and pressure. For that it suffices to place the system in contact with a thermal reservoir or a volume reservoir, respectively. Moreover, the temperature can be measured with a thermometer, and the pressure with a barometer. The question is then, if there are any state functions for which other extremum principles may apply. More precisely, are there any state functions that have a minimum value at equilibrium under conditions that can be controlled in the laboratory? The answer is *yes* and they are called *free energies*. As we shall see, systems hold at constant temperature do not tend to equilibrium states of maximum entropy. Instead they tend to equilibrium states of minimal free energies. To arrive at these novel state functions we have to mathematically recast the dependence of U on two controllable state functions (T and V or T and P), and this is possible through a mathematical operation called *Legendre transform*.

In general, any state function that has a minimum at equilibrium is designated as **thermodynamic potential** by analogy with the potentials of mechanical systems, whose minima correspond to equilibrium (i.e. null force) configurations of the system. Before we introduce the free energies, we discuss another useful thermodynamic potential called enthalpy.

5.2 ENTHALPY

Many natural processes, as well as laboratory experiments, occur at constant pressure. In that case, and since $dP = 0$, heat can be defined as

$$\mathchar'26\mkern-11mu dQ = d(U + PV).$$

Since U, P, and V are all state functions, $U + PV$ is also a state function called **enthalpy** and denoted by the letter H:

$$H \equiv U + PV \qquad\qquad (5.2)$$

The reader can easily check that H has dimensions of energy and the SI unit of H is the joule.

While it is generally true that heat depends on the path chosen to perform a thermodynamic process, it ceases to be so if the pressure is kept constant. Indeed, at constant pressure

$$\mathchar'26\mkern-11mu dQ = dH,$$

or, equivalently

$$Q = \Delta H. \qquad\qquad (5.3)$$

Equation (5.3) motivates the definition of two types of processes:

- **Exothermal** processes are those for which $\Delta H < 0$ as a result of energy being transferred as heat from the system to the surroundings.
- **Endothermal** processes are those for which $\Delta H > 0$ as a result of energy being transferred as heat to the system from the surroundings.

By using equation (5.2) to compute the differential of H, and by considering $dU = \mathchar'26\mkern-11mu dQ - PdV$, it comes that $dH = \mathchar'26\mkern-11mu dQ + V dP$. Since $\mathchar'26\mkern-11mu dQ = CdT$, it is possible to define the heat capacity at constant pressure as a partial derivative of the enthalpy:

$$C_P \equiv \left(\frac{\partial H}{\partial T}\right)_P \qquad\qquad (5.4)$$

Think about it...
What is the enthalpy of an ideal gas composed of N particles?

Answer

For the ideal gas $U = \frac{3}{2}Nk_BT$ and $PV = Nk_BT$. Then, by using equation (5.2)

$$H = \frac{5}{2}Nk_BT,$$

and

$$C_P = \frac{5}{2}Nk_B.$$

In general, C_P will be a function of temperature and therefore, for a change of temperature at constant pressure

$$\Delta H = Q = \int_{T_i}^{T_f} C_P(T)\,dT.$$

It is also interesting to note that for the ideal gas $C_P > C_V$. This relation holds in general because if a fluid it is free to expand and transfer energy as work to the surroundings, more energy transferred as heat is required to increase its temperature than when its volume is kept fixed. In the worked problems we derive a general expression that relates C_P and C_V.

As we show below, systems kept at constant entropy and pressure during a thermodynamic process relax to an equilibrium state where the enthalpy is a minimum. For a process to occur at constant entropy, it should be carried reversibly with no heat exchange. Thus, enthalpy is not often used as an extremum principle. However, it is an important state function which, in view of equation (5.3), can be measured in **calorimetry** experiments. The latter use an apparatus called a calorimeter that measures energy transferred as heat during a physical or chemical reaction. As we will also see, the enthalpy offers a route to determine the Gibbs potential which is of critical importance in Chemistry and Biology.

To establish the extremum principle for the enthalpy we note that

$$đQ = dH - V\,dP \leq T\,dS,$$

whence,

$$dH \leq T\,dS + V\,dP. \tag{5.5}$$

The equality in (5.5) is valid for a reversible process, and establishes the differential of the enthalpy

$$dH = T\,dS + V\,dP \tag{5.6}$$

According to equation (5.6), in a reversible process performed at constant S and P, the system is always in an equilibrium state of constant enthalpy. On the other hand, the inequality in equation (5.5) shows that for all irreversible processes, any change in H will be negative. This means that all irreversible processes will decrease H towards a constant value at equilibrium. Thus the equilibrium state of a system that is kept at constant pressure and constant entropy is a state where enthalpy is a minimum, H_{min} (**enthalpy minimum principle**).

The natural variables of the enthalpy are S and P,

$$H = H(S,P). \tag{5.7}$$

The differential of $H(S,P)$ is

$$dH = \left(\frac{\partial H}{\partial S}\right)_P dS + \left(\frac{\partial H}{\partial P}\right)_S dP,$$

and comparing the latter with equation (5.6) one obtains the **equations of state for the enthalpy**

$$T = \left(\frac{\partial H}{\partial S}\right)_P \text{ and } V = \left(\frac{\partial H}{\partial P}\right)_S,$$

according to which $T = T(S,P)$ and $V = V(S,P)$.

When moving from internal energy $U(S,V)$ to enthalpy $H(S,P)$, the natural variable V is *substituted* by the natural variable P which, as we know, is a partial derivative of the internal energy

$$P = -\left(\frac{\partial U}{\partial V}\right)_S.$$

This observation motivates the introduction of a mathematical tool called the **Legendre transform**, which may be known to the reader from classical mechanics, where it shows up as the operation that relates the Lagrangian L of a mechanical system, with natural variables $(x,\dot{x})$, and the Hamiltonian, with natural variables $(x, p = \frac{\partial L}{\partial \dot{x}})$. The Legendre transform of a function f, in general, allows to change the independent variable from x to the slope $s = df/dx$, preserving a certain symmetry.

LEGENDRE TRANSFORM

The geometric idea behind the Legendre transformation is that for a convex function $f(x)$ there is a one-to-one mapping between the independent variable x and the slope $s = df/dx$, and so the change of variables $x \to s$ is allowed. Instead of an ordinary change of variables, the Legendre transform $g(s)$ of $f(x)$ is defined as

$$g(s) = -f(x(s)) + s\, x(s), \tag{5.8}$$

so that $g(s) + f(x) \equiv s\, x$. From this it follows that $dg/ds = x$, and that $g(s)$ and $f(x)$ are each other's Legendre transforms. The geometrical interpretation of (5.8) is illustrated in Figure 5.2.

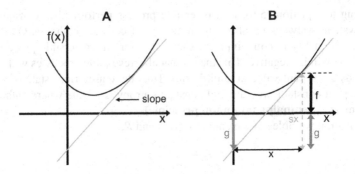

Figure 5.2   The geometrical interpretation of the Legendre transform.

In contrast with classical mechanics, in thermodynamics it is conventional to compute the transform as

$$g(s) = f(x(s)) - s\,x(s), \tag{5.9}$$

and in this case $dg/ds = -x$.

Thus

$$\frac{\partial^2 g}{\partial s^2} = -\frac{\partial x}{\partial s} = -\left(\frac{\partial s}{\partial x}\right)^{-1} = -\left(\frac{\partial^2 f}{\partial x^2}\right)^{-1}, \tag{5.10}$$

which shows that if $f(x)$ is convex ($\partial^2 f/\partial x^2 > 0$), then $g(s)$ will be concave.

For a function of two (or more) variables it is necessary to specify the variable used to perform the Legendre transform. In particular, we define the Legendre transform of a function $f(x,y)$ with regard to variable x as

$$\mathcal{L}_x[f(x,y)] = g(s,y) = f((x(s),y) - sx(s,y) \tag{5.11}$$

In (5.11) s is given by

$$s = \left(\frac{\partial f}{\partial x}\right)_y.$$

A similar definition holds for the Legendre transform with regard to the variable y

$$\mathcal{L}_y[f(x,y)] = g(x,s) = f(x,y(s)) - sy(x,s)$$

with

$$s = \left(\frac{\partial f}{\partial y}\right)_x.$$

When computing the Legendre transforms of thermodynamic potentials, the two variables x, s of the pair thus related are of different types – one of them will be extensive and the other one intensive – and it may be useful to go back and forth from one to the other, depending on the constraints of the system under consideration.

Let us then compute the Legendre transform of the internal energy with respect to the volume. Since $U = U(S,V)$ and

$$P = -\left(\frac{\partial U}{\partial V}\right)_S,$$

then

$$\mathcal{L}_V[U(S,V)] = H(S,P) = U(S,V(S,P)) + PV(S,P). \tag{5.12}$$

By taking into account equations (5.2) and (5.7) we see that (5.12) yields the enthalpy expressed in its natural variables. Note that H, just like U, is a convex function of the extensive variables S and V. However, in light of (5.10), H is a concave function of P.

Equation (5.2) on its own relates the physical quantities involved, and holds irrespectively of the chosen independent variables and mathematical functions. Although there is only one function that expresses the enthalpy as a function of its natural variables, $H = H(S,P)$, defined by (5.12), we will always use the letter H to denote the enthalpy irrespectively of the representation. From a mathematical point of view this stands as a notation abuse, but it is a physically motivated one.

5.3   FREE ENERGY

Here we introduce two important thermodynamic potentials called the Helmholtz free energy (1882), and the Gibbs free energy (1873), named after the 19th century German physiologist and physicist Hermann von Helmholtz (1821–1894) and the contemporaneous American physicist Josiah Willard Gibbs (1839–1903). As we will see, the natural variables of these thermodynamic potentials can be experimentally controlled and measured.

5.3.1   HELMHOLTZ FREE ENERGY

The **Helmholtz free energy**, denoted by the letter F, is the state function defined as

$$F \equiv U - TS \tag{5.13}$$

F has dimensions of energy and the SI unit of F is the joule. As we will see shortly, systems kept at constant volume and temperature during a thermodynamic process relax to an equilibrium state where F is a minimum. Since these conditions are more often found in experiments carried out in physics experiments (e.g. in the study of solids), the Helmholtz free energy is more often used in Physics than in Chemistry.

By computing the differential of F, taking dU as given by the first law, and considering that $dQ \le T dS$, it comes that

$$dF \le -S dT - P dV. \tag{5.14}$$

As in the case of the enthalpy, the equality in (5.14) indicates a reversible process and establishes the differential of the Helmholtz free energy

---

$$dF = -S dT - P dV \tag{5.15}$$

Also, an argument similar to the one we used to analyse the consequences of the inequality in (5.5) leads us to conclude that if T is kept constant and V is fixed, $dF \le 0$, which means that the system will relax to a state of equilibrium ($dF = 0$) where F is a minimum, F_{min} (**Helmholtz free energy minimum principle**).

The natural variables of F are the temperature and the volume,

$$F = F(T, V), \tag{5.16}$$

which can both be easily controlled and measured in laboratory experiments.

Think about it...
Use the Legendre transform operation to identify the natural variables of F

Answer

To move from $U(S,V)$ to $F(T,V)$ one needs to perform the Legendre transform of the internal energy with respect to the entropy:

$$\mathcal{L}_S[U(S,V)] = F(T,V) = U(S(T,V),V) - TS(T,V),$$

with

$$T = \left(\frac{\partial U}{\partial S}\right)_V,$$

showing that the natural variables of F are V and T.

The differential of $F(T,V)$ is

$$dF = \left(\frac{\partial F}{\partial T}\right)_V dT + \left(\frac{\partial F}{\partial V}\right)_T dV,$$

and comparing the latter with (5.6), we obtain the **equations of state for the Helmholtz free energy**

$$S = -\left(\frac{\partial F}{\partial T}\right)_V \text{ and } P = -\left(\frac{\partial F}{\partial V}\right)_S,$$

according to which $S = S(T, V)$ and $P = P(V, S)$, respectively.

From (5.13), it is easy to see that if T is kept constant

$$dF = dU - T dS.$$

But

$$dU - T dS = -P dV.$$

Thus

$$dF = -P dV = \mathchar'26\mkern-12mu dW,$$

or, equivalently

$$\Delta F = -\int_{V_i}^{V_f} P \, dV, \tag{5.17}$$

at constant T.

Thus, if the temperature is kept constant during a thermodynamic process that transfers energy as work, the latter behaves as a state function. In this case the Helmholtz free energy can be interpreted as the maximum (i.e. reversible) work done by the system ($\Delta F < 0$), or the maximum work done on the system ($\Delta F > 0$). For this reason F is sometimes referred as the *work content* of the system, and the letter A (from the german word *arbeit*, which means work) is also used to represent the Helmholtz free energy.

Think about it...

How can one evaluate ΔF for an isothermal compression of the ideal gas?

Answer

By using (5.17) with $P = Nk_B T/V$ one gets

$$\Delta F = -Nk_B T \int_{V_i}^{V_f} \frac{dV}{V} = -Nk_B T \ln\left(\frac{V_f}{V_i}\right) > 0,$$

because $V_f < V_i$.

5.3.2 GIBBS FREE ENERGY

Finally, we introduce the **Gibbs free energy**, another useful state function, denoted by the letter G. There are several equivalent ways to define G. One that is often used is

$$G \equiv H - TS \tag{5.18}$$

Since $H \equiv U + PV$, it is also common to define G as

$$G \equiv U + PV - TS \qquad (5.19)$$

Since $F \equiv U - TS$, G can also be defined as

$$G \equiv F + PV \qquad (5.20)$$

From all the definition above it is easy to see that G also has dimensions of energy, and its SI unit is the joule.

At this stage, it should be easy for the reader to find out that

$$dG \leq -SdT + VdP, \qquad (5.21)$$

from which it follows that the equilibrium state ($dG = 0$) of a system kept under constant temperature and pressure will be the one for which G is a minimum, G_{min} (**Gibbs free energy minimum principle**). The Gibbs free energy is thus a particularly important potential because constant T and P are the easiest constraints to impose on the laboratory as the atmosphere provides them. The vast majority of chemical reactions and biological processes occur under these conditions and this is the reason why the Gibbs potential is the most often used in Chemistry and Biology. Perhaps the most important application of the Gibbs free energy is in the case of phase transitions, for which there will be a dedicated chapter in this book.

As before, the equality in (5.21) refers to a reversible process, and establishes the definition for the differential of the Gibbs free energy

$$dG = -SdT + VdP \qquad (5.22)$$

Since the natural variables of G are T and P

$$G = G(T,P), \qquad (5.23)$$

it comes that

$$dG = \left(\frac{\partial G}{\partial T}\right)_P dT + \left(\frac{\partial G}{\partial P}\right)_T dP.$$

Comparing the last equation with (5.22), it comes that the **equations of state for the Gibbs free energy** are

$$S = -\left(\frac{\partial G}{\partial T}\right)_P \text{ and } V = \left(\frac{\partial G}{\partial P}\right)_T,$$

from which it follows that $S = S(T,P)$ and $V = V(T,P)$.

Think about it...

Identify the natural variables of G by using the Legendre transform

Answer

According to equations (5.18)–(5.20) this can be accomplished in three different ways. Expressed as a function of their natural variables, $H = H(S,P)$ and $G = G(T,P)$. Thus G can be obtained by performing the Legendre transform of H with respect to the entropy:

$$\mathcal{L}_S[H(S,P)] = G(T,P) = H(S(T,P),P) - TS(T,P),$$

with

$$T = \left(\frac{\partial H}{\partial S}\right)_V.$$

Alternatively, since $F = F(T,V)$ and $G = G(T,P)$, the latter can be obtained by performing the Legendre transform of F with respect to the volume:

$$\mathcal{L}_V[F(V,T)] = G(T,P) = F(V(T,P),T) + PV(T,P),$$

with

$$P = -\left(\frac{\partial F}{\partial V}\right)_S.$$

Finally, since $U = U(S,V)$ and $G = G(T,P)$, one can obtain G by performing two Legendre transforms of the internal energy at the same time, one with respect to the volume, and another one with respect to the entropy:

$$\mathcal{L}_{V,S}[U(S,V)] = G(T,P) = U(S(T,P),V(T,P)) + PV(T,P) - TS(T,P),$$

with

$$P = -\left(\frac{\partial U}{\partial V}\right)_S \text{ and } T = \left(\frac{\partial U}{\partial S}\right)_V.$$

Table 5.1 provides a summary of the four thermodynamic potentials, U, H, F, and G, together with S, including their first derivatives.

The complete set of state functions that describe the equilibrium state of a closed single-component system with a fixed number N of particles are U, S, V, T, and P. Only two of them are necessary to completely specify the system's equilibrium state. Expressed as a function of its natural variables, the internal energy contains the complete set of state functions that describe the equilibrium state of such a system. Indeed, $U = U(S,V)$ determines U directly, the equations of state obtained from the differential of U univocally determine the state functions T and P, while S and V are specified by the corresponding constraints. The same is valid for the entropy when expressed in terms of its natural variables (see equations (3.27) and (3.29)), and for the potentials obtained from U through the Legendre transform operation.

Function	Differential	Natural variables	First derivatives
S	$dS = \frac{1}{T}dU + \frac{P}{T}dV$	(U,V)	$\frac{1}{T} = \left(\frac{\partial S}{\partial U}\right)_V, \frac{P}{T} = \left(\frac{\partial S}{\partial V}\right)_U$
U	$dU = TdS - PdV$	(S,V)	$T = \left(\frac{\partial U}{\partial S}\right)_V, P = -\left(\frac{\partial U}{\partial V}\right)_S$
$H = U + PV$	$dH = TdS + VdP$	(S,P)	$T = \left(\frac{\partial H}{\partial S}\right)_P, V = \left(\frac{\partial H}{\partial P}\right)_S$
$F = U - TS$	$dF = -SdT - PdV$	(T,V)	$S = -\left(\frac{\partial F}{\partial T}\right)_V, P = -\left(\frac{\partial F}{\partial V}\right)_T$
$G = H - TS$	$dG = -SdT + VdP$	(T,P)	$S = -\left(\frac{\partial G}{\partial T}\right)_P, V = \left(\frac{\partial G}{\partial P}\right)_T$

Table 5.1
Differentials, natural variables and first derivatives of the entropy and thermodynamic potentials for a closed system.

Additional functions such as $U(P,V)$, $U(S,P)$ etc. do not *uniquely* specify the equilibrium state of a system. Imagine that we had an expression for the function $U(S,P)$. Then, instead of taking the derivatives of $U(S,V)$ to obtain the state functions T and P, we would have to integrate the equation

$$P = -\left(\frac{\partial U}{\partial V}\right)_S$$

to obtain the volume V, which requires an integration constant. In other words, we would need to have a value for U at some reference volume V. Thus, $U(S,P)$ does not provide a complete description of the thermodynamic system. This is, in fact, another reason why S and V are the natural variables of U. Other functions such as $U(T,V)$, $S(T,V)$ or $H(T,P)$ are not associated with extremum principles. They are, however, useful being components of $F(T,V)$ and $G(T,P)$.

Think about it...
Why should the potential G be considered a free energy?

Answer

To answer this question we need to consider the more general situation in which the system performs expansion work (i.e. $-PdV$), as well as at least another form of work that we represent by dW'. The first law then reads $dU = dQ + dW$, with $dW = dW' - PdV$, and the fundamental constraint is $dU = TdS + dW$. Consider $G = F + PV$. If P is constant and T is constant as well, it is easy to see that $dG = dW'$. Thus, dG represents the maximum work done by (or on the) system during a reversible thermodynamic process, except the volume work. Note that in the more general case that dW' is considered, the differential of F is $dF = -SdT - PdV + dW'$. Thus, if T is constant dF represents the maximum reversible work done by (or on the) system or the system, including expansion work.

5.4   MAXWELL RELATIONS

Since U, H, F, and G are state functions, their differential equations are exact differentials. This observation has an important consequence, which is the establishment of the so-called **Maxwell Relations** that we now discuss.

Consider, for example, the exact differential of the internal energy

$$dU = TdS - PdV,$$

with

$$T = \left(\frac{\partial U}{\partial S}\right)_V \text{ and } P = -\left(\frac{\partial U}{\partial V}\right)_S.$$

Since dU is exact

$$\frac{\partial^2 U}{\partial V \partial S} = \frac{\partial^2 U}{\partial S \partial V}.$$

Thus

$$\left(\frac{\partial T}{\partial V}\right)_S = -\left(\frac{\partial P}{\partial S}\right)_V. \tag{5.24}$$

Equation (5.24) is an example of a Maxwell relation.

Since dH, dF, and dG are also exact differentials, the following Maxwell relations follow:

$$\left(\frac{\partial T}{\partial P}\right)_S = \left(\frac{\partial V}{\partial S}\right)_P, \tag{5.25}$$

$$\left(\frac{\partial S}{\partial V}\right)_T = \left(\frac{\partial P}{\partial T}\right)_V, \tag{5.26}$$

and

$$\left(\frac{\partial S}{\partial P}\right)_T = -\left(\frac{\partial V}{\partial T}\right)_P. \tag{5.27}$$

The Maxwell relations are not exhausted by equations (5.24)–(5.27). Indeed, as we will see shortly, other Maxwell relations can be established for open systems, and for other thermodynamic systems with different work functions. In Chapter 8, we will study in detail the case of magnetic work.

Partial derivatives are not just mathematical expressions. They tell us how the dependent variable (in the numerator) changes when the independent variable (in the denominator) is varied in a process where some other state function is kept constant. The last two Maxwell relations are particularly interesting because they give quantities one cannot measure from measurable ones. Consider the case of equation (5.26), where the state functions on the right hand side are all easily measured in the laboratory. The equality expressed in (5.26) tells us that the fractional change in pressure (the dependent variable) with temperature (the independent one) when the volume is kept fixed, is equal to the fractional change in entropy with volume when the temperature is kept constant. The negative sign on equation (5.27) is also quite informative. Indeed, since the volume is expected to increase with temperature at constant pressure, the minus sign means that the entropy will decrease with increasing pressure at constant temperature.

5.5　THERMODYNAMIC COEFFICIENTS

In thermodynamics, a **generalised force** is any property that can be defined as a partial derivative of the internal energy with regard to any other variable. T and P are examples of generalised forces. Another generalised force is the **internal pressure**

$$\pi_T \equiv \left(\frac{\partial U}{\partial V} \right)_T, \tag{5.28}$$

which is associated with the existence of intermolecular interactions between the system's particles. In the case of the ideal gas, where no such interactions exist, and $U = U(T)$, $\pi_T = 0$.

On the other hand, a **generalised susceptibility** is an entity that quantifies the degree of variation of a thermodynamic property resulting from a generalised force. Since a generalised susceptibility quantifies the response of a property as a result of changing another property it is also termed of **response function**. Two important examples are the **coefficient of isobaric expansivity**

$$\beta_P \equiv \frac{1}{V} \left(\frac{\partial V}{\partial T} \right)_P, \tag{5.29}$$

and the **coefficient of adiabatic expansivity**

$$\beta_S \equiv \frac{1}{V} \left(\frac{\partial V}{\partial T} \right)_S. \tag{5.30}$$

β_P and β_S measure the fractional variation of the volume with temperature (i.e. the amount of system expansion) when P and S are kept constant, respectively. β_P is also designated by the **expansion coefficient** and represented by the greek letter α. The reader can easily check that $\beta_P = 1/T$ in the case of the ideal gas. In general, β_P is positive for fluids, except in the case of water between $0°C - 4°C$.

The **thermal pressure coefficient**, γ_V, quantifies the variation of entropy with volume at constant temperature

$$\gamma_V \equiv \left(\frac{\partial S}{\partial V} \right)_T = \left(\frac{\partial P}{\partial T} \right)_V, \tag{5.31}$$

where we used Maxwell relation (5.26) in the second equality. For the ideal gas $\gamma_V = Nk_B/V$. In general, the pressure is expected to increase with temperature at fixed V. Thus, the entropy should increase with volume at constant temperature.

To quantify the fractional variation of volume with pressure (i.e. the level of system compression) there are two entities. One is called the **coefficient of adiabatic compressibility**

$$k_S \equiv -\frac{1}{V}\left(\frac{\partial V}{\partial P}\right)_S, \tag{5.32}$$

and the other is the **coefficient of isothermal compressibility**

$$k_T \equiv -\frac{1}{V}\left(\frac{\partial V}{\partial P}\right)_T. \tag{5.33}$$

Since the volume is expected to decrease with pressure at constant temperature, k_T is defined with a minus sign in order to be a positive quantity. For the ideal gas $k_T = 1/P$. For solids and some liquids $k_T \approx 0$, while gases present much higher values of k_T as a result of being much more compressible.

A particularly interesting coefficient, which is often used to provide a measure of how a real gas deviates from the ideal gas behaviour, is the so-called **Joule-Thomson coefficient**

$$\mu_{JT} \equiv \left(\frac{\partial T}{\partial P}\right)_H. \tag{5.34}$$

The Joule-Thomson coefficient measures how temperature changes with pressure during an isenthalpic thermodynamic process, i.e., one that occurs at fixed enthalpy. This type of process is often called a **throttling process**. To perform the measurement one uses an apparatus that consists of a vessel with a porous wall that separates two parts, each one tapped by a piston (Figure 5.3). The pistons and the walls of the vessel are made of a thermally insulating material, so that no energy is transferred as heat, during a thermodynamic process. The gas whose Joule-Thomson coefficient one wishes to measure is placed within the two parts, say one mole of gas is placed in part 1 with P_1, V_1, T_1, and another quantity is placed in part 2 with $P_2 < P_1, V_2, T_2$. Since there is a pressure difference, the gas on part 1, which is at $P_1 > P_2$, will pass

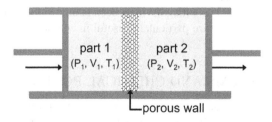

Figure 5.3   An experimental setup to measure the Joule-Thomson coefficient of a gas. The vessel composed of two parts separated by a porous wall, as well as the two pistons that tap the two parts are composed of adiabatic materials. The porous wall allows for gas to pass slowly from 1 to 2.

slowly through the porous wall towards part 2. To keep the pressure of both compartments constant the pistons are moved slowly in the direction that is consistent with

$P_1 > P_2$. To see that this corresponds to a process at fixed enthalpy it suffices to note the following. The variation in internal energy associated with the passage of the gas is given by $\Delta U = U_2 - U_1$. On the other hand, the work performed by the gas during the process is $W = P_1 V_1 - P_2 V_2$, because in part 2 the gas is compressed, while in part 1 it is expanded. Since there is no energy transferred as heat, the first law states that

$$U_2 - U_1 = P_1 V_1 - P_2 V_2,$$

which is equivalent to

$$H_2 = H_1,$$

because $H = U + PV$. Since the gas in each compartment is at temperatures T_1 and T_2, in order to measure μ_{JT} is suffices to note that

$$\mu_{JT} = \left(\frac{\partial T}{\partial P}\right)_H = \lim_{T_2 \to T_1} \left(\frac{\Delta T}{\Delta P}\right),$$

with $\Delta P = P_2 - P_1$ and $\Delta T = T_2 - T_1$.

The use of Maxwell relations in combination with the properties of partial derivatives can be extraordinarily useful while manipulating the thermodynamic coefficients and establishing relations between them, and between them and other entities such as the heat capacities. In particular, it is possible to show (see problem 5.2) that

$$\left(\frac{\partial T}{\partial P}\right)_H = \frac{T\left(\frac{\partial V}{\partial T}\right)_P - V}{C_P},$$

which is zero for the ideal gas. Thus, for one mole of ideal gas μ_{JT} is actually zero. Real gases have negative μ_{JT} at high temperatures (the gas cools on expansion) and positive μ_{JT} at low temperatures (the gas heats on expansion). The temperature regime in which μ_{JT} passes through zero, approaching the ideal gas behaviour, is termed the **inversion temperature**.

Several thermodynamic coefficients have been tabulated for many substances. In the case of solids the latter are particularly useful in engineering, materials science and related areas.

5.6   OPEN SYSTEMS AND CHEMICAL POTENTIAL

So far, we have focused on thermodynamic processes for which N is constant. In other words, we have been developing the theory of thermodynamics for closed systems. For these systems we found that $U = U(S,V)$, or equivalently, that $S = S(U,V)$. However, we need to bear in mind that the number of particles is not conserved in many physical processes. Clearly, if the number of particles that composes the thermodynamic system is allowed to change, then the system's internal energy will change as well. Thinking of an ideal gas, where the particles have only kinetic energy, it is clear that if particles enter or leave the system, the system's internal energy will change. The same reasoning applies in the case of entropy, as the number of microstates depends on the number of particles. Therefore, in general, the internal energy and the entropy depend on the number of particles that compose the thermodynamic system. In particular, if the system only contains one type of particle

$$U = U(S,V,N),$$

(5.35)

and

$$S = S(U,V,N).$$

(5.36)

Equation (5.35) is termed the **fundamental equation in the energy representation**, while (5.36) is the **fundamental equation in the entropy representation**.

From (5.35) the differential of U is

$$dU = \left(\frac{\partial U}{\partial S}\right)_{V,N} dS + \left(\frac{\partial U}{\partial V}\right)_{S,N} dV + \left(\frac{\partial U}{\partial N}\right)_{S,V} dN,$$

where the last partial derivative defines a novel intensive state function called **chemical potential**:

$$\mu = \left(\frac{\partial U}{\partial N}\right)_{S,V}$$

(5.37)

The concept of chemical potential was introduced by Josiah Willard Gibbs in his famous work *On the Equilibrium of Heterogeneous Substances* (1874) that forms the basis of modern physical chemistry and of the thermodynamics of phase transitions. The chemical potential represents the change in internal energy when one particle is transferred to, or from, the system. The SI unit of μ is J mol^{-1}.

The designation chemical potential is perhaps a misnomer as it may suggest that a change in the number of particles necessarily requires the occurrence of a chemical reaction, which is definitely not the case. A familiar situation is that of phase transitions in which there is transfer of particles between different phases in the absence of chemical reactions.

As we will discuss later in this chapter, the chemical potential is related to the number of particles in much the same way that the temperature is related to the internal energy, and the pressure is related to the volume. However, since we do not have a direct way to measure the chemical potential (as we do for the temperature and pressure) it is more difficult to develop an intuition for its meaning.

By taking into consideration the exposed above, a more general form of the fundamental constraint is

$$dU = TdS - PdV + \mu dN$$

(5.38)

In equation (5.38)

$$T = \left(\frac{\partial U}{\partial S}\right)_{V,N},$$

$$P = -\left(\frac{\partial U}{\partial V}\right)_{S,N},$$

and

$$\mu = \left(\frac{\partial U}{\partial N}\right)_{S,V}.$$

Note that the partial derivatives above are taken with the additional constraint N.

The fundamental equation contains the complete set of thermodynamic information of a single-component system that exchanges particles with the surroundings, namely, U, T, P, μ, N, S, and V. Indeed, (5.35) gives U directly, the differential of U gives T, P, and μ, and V and N are specified by the constraints.

Taking into account that dU is exact, it comes that

$$\frac{\partial^2 U}{\partial N \partial S} = \frac{\partial^2 U}{\partial S \partial N},$$

and

$$\frac{\partial^2 U}{\partial N \partial V} = \frac{\partial^2 U}{\partial V \partial N}.$$

These equations define two additional Maxwell relations for the internal energy:

$$\left(\frac{\partial T}{\partial N}\right)_{S,V} = \left(\frac{\partial \mu}{\partial S}\right)_{V,N}, \tag{5.39}$$

and

$$-\left(\frac{\partial P}{\partial N}\right)_{S,V} = \left(\frac{\partial \mu}{\partial V}\right)_{S,N}. \tag{5.40}$$

For an open system composed by one type of particles it is easy to see that the differential form of the fundamental equation in the entropy representation is

$$dS = \left(\frac{1}{T}\right)dU + \left(\frac{P}{T}\right)dV - \left(\frac{\mu}{T}\right)dN \tag{5.41}$$

The fundamental equation in the entropy representation also contains the complete set of thermodynamic information of a single-component system that exchanges particles with the surroundings. The two formulations expressed by (5.35) and (5.36) are therefore equivalent.

Think about it...
Show that

$$\left(\frac{\partial S}{\partial N}\right)_{U,V} = -\frac{\mu}{T}.$$

Answer

The differential of $S = S(U,V,N)$ is

$$ds = \left(\frac{\partial S}{\partial U}\right)_{V,N} dU + \left(\frac{\partial S}{\partial V}\right)_{U,N} dV + \left(\frac{\partial S}{\partial N}\right)_{U,V} dN$$

$$= \frac{1}{T} dU + \frac{P}{T} dV + \left(\frac{\partial S}{\partial N}\right)_{U,V} dN.$$

Using the fundamental constraint for dU (5.38), the result follows immediately.

5.7   MULTICOMPONENT SYSTEMS

If the system contains particles of m different types:

$$U = U(T,S,V,N_1,...,N_m), \tag{5.42}$$

and

$$S = S(T,U,V,N_1,...,N_m). \tag{5.43}$$

The differential of (5.42) is

$$dU = TdS - PdV + \sum_{i=1}^{m} \mu_i dN_i \tag{5.44}$$

In equation (5.44), μ_i is the **chemical potential of species** i:

$$\mu_i = \left(\frac{\partial U}{\partial N_i}\right)_{S,V,\{N_{j\neq i}\}}. \tag{5.45}$$

The differential of (5.43) is

$$dS = \left(\frac{1}{T}\right)dU + \left(\frac{P}{T}\right)dV - \sum_{i=1}^{m} \left(\frac{\mu_i}{T}\right)dN_i \tag{5.46}$$

In equation (5.46)

$$-\left(\frac{\mu_i}{T}\right) = \left(\frac{\partial S}{\partial N_i}\right)_{U,V,\{N_{j\neq i}\}}. \tag{5.47}$$

Equations (5.44) and (5.46) are the most general differential forms of the fundamental constraint in the internal energy representation and in the entropy representation, respectively.

Since the thermodynamic potentials H, F, and G are obtained from the internal energy, they must also depend on the number of particles that compose the system. In the more general case of an open system containing particles of m different types we have:

$$H = H(S, P, N_1, N_2, ..., N_m),$$

$$F = F(T, V, N_1, N_2, ..., N_m),$$

and

$$G = G(T, P, N_1, N_2, ..., N_m).$$

The differential equations for H, F, and G are thus given by

$$dH = TdS + VdP + \sum_{i=1}^{m} \mu_i dN_i, \text{ with } \mu_i = \left(\frac{\partial H}{\partial N_i}\right)_{S,P,\{N_{j\neq i}\}}, \tag{5.48}$$

$$dF = -SdT - PdV + \sum_{i=1}^{m} \mu_i dN_i, \text{ with } \mu_i = \left(\frac{\partial F}{\partial N_i}\right)_{T,V,\{N_{j\neq i}\}}, \tag{5.49}$$

and

$$dG = -SdT + VdP + \sum_{i=1}^{m} \mu_i dN_i, \text{ with } \mu_i = \left(\frac{\partial G}{\partial N_i}\right)_{T,P,\{N_{j\neq i}\}}, \tag{5.50}$$

with

$$\mu_i = \left(\frac{\partial U}{\partial N_i}\right)_{S,V,\{N_{j\neq i}\}} = \left(\frac{\partial H}{\partial N_i}\right)_{S,P,\{N_{j\neq i}\}} = \left(\frac{\partial F}{\partial N_i}\right)_{T,V,\{N_{j\neq i}\}} = \left(\frac{\partial G}{\partial N_i}\right)_{T,P,\{N_{j\neq i}\}}.$$

The equalities above indicate that there are several alternative methods to determine the chemical potential. According to the last term, one can determine μ_i by adding N_i particles to the system while keeping T, P, and the number of the other particles constant. The corresponding change in G, dG, to dN_i (in the limit as $dN_i \to 0$), then measures μ_i.

5.8   THE GRAND POTENTIAL

Another useful thermodynamic potential, is the so-called **grand potential**. The grand potential also goes by the name of **Landau potential**, and is generally represented by the Greek letter Ω. The grand potential is the Legendre transform of the Helmholtz free energy with regard to the number of particles

$$\mathcal{L}_N[F(T,V,N)] = F(T,V,N(V,T,\mu)) - \mu N(T,V,\mu), \tag{5.51}$$

with

$$\mu = \left(\frac{\partial F}{\partial N}\right)_{T,V}.$$

The natural variables of Ω are T, V, and μ:

$$\Omega = \Omega(T, V, \mu).$$

In Chapter 6 we will derive a very important relation between G and μ according to which

$$\mu = \frac{G}{N}.$$

Using this result in (5.51), it comes that

$$\Omega = F - G.$$

Since $G = F + PV$, we can finally write

$$\Omega = -PV \tag{5.52}$$

The grand potential, as defined by (5.52), turns out to be a very useful equation to calculate the equation of state of various physical systems, especially in the context of statistical physics.

5.9   MASSIEU FUNCTIONS

By considering the fundamental relation of thermodynamics in the internal energy representation $U = U(S,V,N)$, we were able to define four thermodynamic potentials, namely H, F, G, and Ω, by means of the Legendre transform. Since thermodynamics can be formulated in the entropy representation $S = S(U,V,N)$, it is natural to ask what are the Legendre transforms of the entropy, and if they share the same interest and applicability of the thermodynamic potentials. The Legendre transforms of the entropy are designated by **Massieu functions** or **free entropies**, being particularly useful in statistical mechanics, in the theory of irreversible thermodynamics, and in the theory of fluctuations. In general, these functions do not have any special names, but there are two exceptions. One, the so-called **Helmholtz free entropy** and denoted by the Greek letter Φ, is the Legendre transform of S with respect to the internal energy

$$\mathcal{L}_U[S(U,V,N)] = S(U(1/T),V,N) - \frac{1}{T}U(1/T,V,N),$$

with

$$\frac{1}{T} = \left(\frac{\partial S}{\partial U}\right)_{V,N}.$$

The natural variables of Φ are $1/T, V$ and N

$$\Phi = \Phi(1/T,V,N).$$

The reason why Φ is called Helmholtz free entropy is simply because

$$\Phi = \frac{-U + TS}{T} = \frac{-F}{T}.$$

The second Massieu function that is entitled to have a specific name is the **Gibbs free entropy**. It is denoted by the Greek letter Ξ, and is the Legendre transform of the entropy with respect to both the internal energy and volume

$$\mathcal{L}_{U,V}[S(U,V,N)] = S(U(1/T),(P/T),N) - \frac{1}{T}U(1/T,P/T,N) - \frac{P}{T}V(1/T,P/T,N)$$

with

$$\frac{1}{T} = \left(\frac{\partial S}{\partial U}\right)_{V,N} \text{ and } \frac{P}{T} = \left(\frac{\partial S}{\partial V}\right)_{U,N}.$$

Note that

$$\Xi = \frac{-U - TS - PV}{T} = \frac{-G}{T},$$

which justifies the name Gibbs free entropy.

5.10 THERMODYNAMIC EQUILIBRIUM REVISITED

Let us consider a system that is composed of two parts (A and B), separated by an imaginary boundary. The latter behaves as a movable diathermal wall that is also permeable allowing particles to be transferred between A and B.

The system is isolated, which means that it does not change energy, volume or particles with the surroundings (Figure 5.4). Likewise, the total internal energy is fixed

$$U = U_A + U_B = \text{constant}, \tag{5.53}$$

the total volume is fixed

$$V = V_A + V_B = \text{constant}, \tag{5.54}$$

and the total number of particles is also fixed

$$N = N_A + N_B = \text{constant}. \tag{5.55}$$

Equations (5.53)–(5.55) are often termed **conservation** equations.

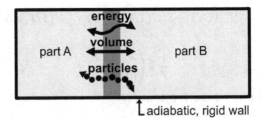

Figure 5.4   A system composed of two (internal) parts separated by a diathermic, permeable, and movable wall. The system is isolated from the surroundings by an adiabatic, impermeable, rigid wall.

According to the maximum entropy principle $S = S_A + S = S_A + S_B$ is maximum at equilibrium. Therefore

$$dS = dS_A + dS_B = 0.$$

Considering that $S_A = S_A(U_A, V_A, N_A)$ and $S_B = S_B(U_B, V_B, N_B)$, the differential of S is

$$dS = \left(\frac{\partial S_A}{\partial U_A}\right)_{V_A,N_A} dU_A + \left(\frac{\partial S_A}{\partial V_A}\right)_{U_A,N_A} dV_A + \left(\frac{\partial S_A}{\partial N_A}\right)_{U_A,V_A} dN_A$$

$$+ \left(\frac{\partial S_B}{\partial U_B}\right)_{V_B,N_B} dU_B + \left(\frac{\partial S_B}{\partial V_B}\right)_{U_B,N_B} dV_B + \left(\frac{\partial S_B}{\partial N_B}\right)_{U_B,V_B} dN_B,$$

Since

$$dU = 0 \Rightarrow dU_A = -dU_B$$
$$dV = 0 \Rightarrow dV_A = -dV_B$$
$$dN = 0 \Rightarrow dN_A = -dN_B,$$

we can rewrite dS as

$$dS = \left[\left(\frac{\partial S_A}{\partial U_A}\right)_{V_A,N_A} - \left(\frac{\partial S_B}{\partial U_B}\right)_{V_B,N_B}\right] dU_A$$

$$+ \left[\left(\frac{\partial S_A}{\partial V_A}\right)_{U_A,N_A} - \left(\frac{\partial S_B}{\partial V_B}\right)_{U_B,N_B}\right] dV_A \qquad (5.56)$$

$$+ \left[\left(\frac{\partial S_A}{\partial N_A}\right)_{U_A,V_A} - \left(\frac{\partial S_B}{\partial N_B}\right)_{U_B,V_B}\right] dN_A.$$

Because $dS = 0$ in equilibrium, for arbitrary and independent values of dU_A, dV_A, and dN_A, we must have

$$\left(\frac{\partial S_A}{\partial U_A}\right)_{V_A,N_A} = \left(\frac{\partial S_B}{\partial U_B}\right)_{V_B,N_B}, \qquad (5.57)$$

$$\left(\frac{\partial S_A}{\partial V_A}\right)_{U_A,N_A} = \left(\frac{\partial S_B}{\partial V_B}\right)_{U_B,N_B}, \qquad (5.58)$$

$$\left(\frac{\partial S_A}{\partial N_A}\right)_{U_A,V_A} = \left(\frac{\partial S_B}{\partial N_B}\right)_{U_B,V_B}. \qquad (5.59)$$

Equations (5.57)–(5.59) are equivalent to

$$\frac{1}{T_A} = \frac{1}{T_B}, \qquad (5.60)$$

$$\frac{P_A}{T_A} = \frac{P_B}{T_B}, \qquad (5.61)$$

and

$$\frac{\mu_A}{T_A} = \frac{\mu_B}{T_B}. \qquad (5.62)$$

Equation (5.60) establishes the condition for **thermal equilibrium**, and is equivalent to $T_A = T_B$. We decided to keep the fractional expressions to emphasise the fact that they were derived from the fundamental equation in the entropy representation. Since $T_A = T_B$, equation (5.61) establishes the condition for **mechanical equilibrium**, which occurs when the pressures of the two parts are equal

$$P_A = P_B.$$

Finally, equation (5.62) establishes the condition for **chemical equilibrium**, which occurs when the chemical potential of the two parts are equal

$$\mu_A = \mu_B.$$

A system that satisfies the three equilibrium conditions is said to be in **thermodynamic equilibrium**.

To get a more physical intuition for the chemical potential, let us consider the situation in which the parts of the system do not exchange energy (because $T_A = T_B = T$), and do not exchange volume either (because $P_A = P_B = P$). In this case equation (5.56) resumes to

$$dS = \left[\frac{\mu_B - \mu_A}{T} \right] dN_A.$$

So, if $\mu_B > \mu_A$, dN_A must be positive to ensure that the entropy will be maximum at equilibrium. This means that particles will move from part B to part A until $\mu_A = \mu_B$. Alternatively, if $\mu_B < \mu_A$, dN_A must be negative, which means that particles will move from part A to part B, until $\mu_A = \mu_B$.

In approaching equilibrium, particles flow from parts with high chemical potential to parts with low chemical potential

---

So, just like energy must change to equalise the temperature, and volume must change to equalise the pressure, the number of particles must change to equalise the chemical potential.

Think about it...
If instead of the equilibrium condition for the entropy one uses the equilibrium condition for G, does one arrive at equations (5.60)–(5.62)?

Answer

The conditions for thermal and mechanical equilibrium are automatically satisfied because $dG = 0$ when both T and P are fixed. To see if the condition for chemical equilibrium is satisfied we need to consider that $G_A + G_B = G$ and, therefore, $dG_A + dG_B = 0$. Interpreting the index i in equation (5.50) as identifying the particles in part A and the particles in part B one can write that

$$dG = \mu_A dN_A + \mu_B dN_B = 0.$$

Since $dN_A = -dN_B$, it follows that $\mu_A = \mu_B$. Thus, G yields the same result as S.

The reader may confirm that H and F lead to the same conditions for internal equilibrium.

5.11   STABILITY OF EQUILIBRIUM STATES

Due to the molecular motions occurring at the microscopic scale, intensive properties continuously fluctuate (i.e. exhibit very small changes around their equilibrium values) (Figure 1.2), causing the corresponding extensive properties to fluctuate locally. Since the equilibrium state of an isolated thermodynamic system is the one for which the entropy is maximum, any fluctuation of the volume, internal energy, number of particles, or any combination of the former, can only reduce the system's entropy. If, in response to a fluctuation, the system spontaneously goes back to its original equilibrium state, the system is considered **intrinsically stable**.

Let us consider a differential fluctuation of the internal energy, volume, and number of particles from their equilibrium values such that $U = U_{eq} + dU$, $V = V_{eq} + dV$, and $N = N_{eq} + dN$, respectively, and expand the entropy, $S = S(U,V,N)$, in a Taylor series around the equilibrium point:

$$\Delta S = dS + \frac{1}{2}d^2S + ..., \tag{5.63}$$

where $\Delta S = S - S_{eq}$, the term dS represents the first-order term containing dU, dV, and dN and is given by equation (5.41), while d^2S represents the second-order terms, and so on.

Since dS is zero at equilibrium, the leading term deviating S from S_{eq} is the second-order term d^2S. From a mathematical standpoint, the maximum entropy principle implies, not only that $dS = 0$, but also that the **second-order variation of** S (also termed second-order response) is negative (Figure 5.1 A)

$$d^2S < 0, \tag{5.64}$$

since the surface $S(U,V,N)$ is concave around a maximum (Figure 5.1 A).

An equation equivalent to (5.63) can be written for the internal energy, U. In this case, the principle of minimum energy implies that $dU = 0$, and that the **second-order variation of** U is positive

$$d^2U > 0, \tag{5.65}$$

since the surface $U(S,V,N)$ is convex around a minimum (Figure 5.1 B). When another thermodynamic potential (always expressed as a function of its natural variables) is considered, a condition equivalent to (5.65) applies. Namely, $d^2H > 0$, $d^2F > 0$, or $d^2G > 0$.

The condition that S is a maximum or, alternatively, that U (or H, F or G) is a minimum, is the requirement of **stability** of the corresponding equilibrium states. In what follows we analyse the consequences of stability of thermodynamic states with regard to small fluctuations, namely, on the sign of certain partial derivatives, and on the sign of thermodynamic coefficients.

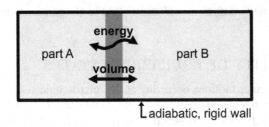

Figure 5.5   A system composed of two (internal) parts separated by a diathermic and movable wall that can be real or imaginary. The system is isolated from the surroundings by an adiabatic, rigid wall.

In the previous section we explored the consequences of $dS = 0$ for an isolated system in equilibrium composed of two parts. Here we explore the consequences that $d^2S < 0$ and $d^2U > 0$.

In order to do so, let us go back to the composite system of the previous section, but this time we will consider a simpler situation in which the system can still exchange energy and volume between its parts, but no particles move from one part to the other because the boundary is impermeable (Figure 5.5). In this case

$$dS = \left(\frac{\partial S_A}{\partial U_A}\right)_{V_A,N_A} dU_A + \left(\frac{\partial S_A}{\partial V_A}\right)_{U_A,N_A} dV_A$$
$$+ \left(\frac{\partial S_B}{\partial U_B}\right)_{V_B,N_B} dU_B + \left(\frac{\partial S_B}{\partial V_B}\right)_{U_B,N_B} dV_B.$$

The second-order variation in S is given by

$$d^2S = S_A^{uu}(dU_A)^2 + 2S_A^{uv} dU_A dV_A + S_A^{vv}(dV_A)^2$$
$$+ S_B^{uu}(dU_B)^2 + 2S_B^{uv} dU_B dV_B + S_B^{vv}(dV_B)^2, \tag{5.66}$$

where the notation S^{xy} is used to represent the second derivative $\partial^2 S/\partial X \partial Y$. This equation takes into account *all possible variations* in S, resulting from energy exchange, volume exchange, or both energy and volume exchange between two parts. Using the conservation equations (5.53) and (5.54), and the fact that in equilibrium the intensive properties T (5.57) and P (5.58) are the same in parts A and B, the condition expressed by (5.64) requires that

$$\frac{1}{2}\left[S^{uu}(dU)^2 + 2S^{uv}dUdV + S^{vv}(dV)^2\right] < 0, \tag{5.67}$$

where we dropped the subscripts A and B because we are only interested in the sign of d^2S. In the context of linear algebra the left-hand side of equation (5.67) is termed a **quadratic form**.

For infinitesimal changes in U with V fixed, equation (5.64) requires that $S^{uu} < 0$. Alternatively, for infinitesimal changes in V with U fixed, it is necessary that $S^{vv} < 0$. However, for all possible changes in U and V a more general condition can be established, which ensures that $d^2S < 0$.

The quadratic form on the left-hand side of (5.67) can be written as

$$d^2S = \left[dU\,dV\right]\mathbf{S}\begin{bmatrix}dU \\ dV\end{bmatrix},$$

with $\mathbf{S}$ being the **symmetric Hessian matrix**:

$$\mathbf{S} = \begin{bmatrix}S_{uu} & S_{uv} \\ S_{vu} & S_{vv}\end{bmatrix}.$$

For d^2S to be negative, $\mathbf{S}$ must be **negative definite** for all possible variations of U and V. It is a necessary and sufficient condition for $\mathbf{S}$ to be negative definite that:

$$S_{uu} < 0, \tag{5.68}$$

and

$$\begin{vmatrix}S_{uu} & S_{uv} \\ S_{uv} & S_{vv}\end{vmatrix} = S_{uu}S_{vv} - S_{uv}S_{uv} > 0 \tag{5.69}$$

Equations (5.68) and (5.69) **determine the stability criteria for the equilibrium state**. A condition of the type given by (5.69) is designated by Callen as *fluting condition*. When both equations are satisfied, the equilibrium state is **locally stable**, i.e., it is stable with regard to differential changes of U and V. They insure locally that the surface S will not lie above its local tangent plane.

If, instead of considering a variation in S, we consider a variation in the internal energy U, we obtain an equation equivalent to (5.67)

$$\frac{1}{2}\left[U^{ss}(dS)^2 + 2U^{sv}dSdV + S^{vv}(dV)^2\right] > 0. \tag{5.70}$$

For d^2U to be positive, the Hessian matrix

$$\mathbf{U} = \begin{bmatrix}U_{ss} & U_{sv} \\ U_{sv} & U_{vv}\end{bmatrix} = U_{ss}U_{vv} - U_{sv}U_{sv},$$

must be **positive definite**.

Thus, it is necessary that

$$U_{ss} > 0, \tag{5.71}$$

and

$$U_{ss}U_{vv} - U_{sv}U_{sv} > 0. \tag{5.72}$$

Equation (5.72) is the *fluting condition* for the internal energy. Equations (5.71) and (5.72) determine the stability criteria for the internal energy.

Think about it...

Use the stability conditions for the internal energy to show that $\left(\frac{\partial T}{\partial S}\right)_{V,N} > 0$ and $\left(\frac{\partial P}{\partial V}\right)_{S,N} < 0$.

Answer

The first inequality follows from $U_{ss} > 0$ and the second from $U_{vv} > 0$:

$$U_{ss} = \left(\frac{\partial^2 U}{\partial S^2}\right)_{V,N} = \frac{\partial}{\partial S}\left(\frac{\partial U}{\partial S}\right)_{V,N} = \left(\frac{\partial T}{\partial S}\right)_{V,N},$$

and

$$U_{vv} = \left(\frac{\partial^2 U}{\partial V^2}\right)_{S,N} = \frac{\partial}{\partial V}\left(\frac{\partial U}{\partial V}\right)_{S,N} = -\left(\frac{\partial P}{\partial V}\right)_{S,N},$$

where we used the thermodynamic definitions of temperature and pressure. Thus

$$U_{ss} > 0 \quad \rightarrow \quad \left(\frac{\partial T}{\partial S}\right)_{V,N} > 0,$$

and

$$U_{vv} > 0 \quad \rightarrow \quad \left(\frac{\partial P}{\partial V}\right)_{S,N} < 0.$$

5.11.1 STABILITY AND THE SIGN OF THERMODYNAMIC COEFFICIENTS

The stability criteria for the entropy and for internal energy determine the sign of thermodynamic coefficients. To illustrate this point we explore the consequences of equations (5.68) and (5.69). We start by evaluating S_{uu}:

$$S_{uu} = \left(\frac{\partial^2 S}{\partial U^2}\right)_V = \frac{\partial}{\partial U}\left(\frac{1}{T}\right)_V = -\frac{1}{T^2}\left(\frac{\partial T}{\partial U}\right)_V = -\frac{1}{T^2 C_V},$$

where we used the reciprocal rule, and C_V given by (2.15). To satisfy the stability criterion established by (5.68) it is necessary that

$$C_V > 0 \qquad (5.73)$$

The inequality expressed by equation (5.73) determines the criterion for **thermal stability**: the internal energy of a system in a well-defined equilibrium state must always increase in response to a isochoric fluctuation that increases the system's temperature.

Let us now evaluate S_{uv}:

$$S_{uv} = \frac{\partial}{\partial V}\left[\left(\frac{\partial S}{\partial U}\right)_V\right]_U = \frac{\partial}{\partial V}\left(\frac{1}{T}\right)_U$$

$$= -\frac{1}{T^2}\left(\frac{\partial T}{\partial V}\right)_U$$

$$= \frac{1}{T^2}\frac{(\partial U/\partial V)_T}{(\partial U/\partial T)_V} = \frac{T\gamma_V - P}{T^2 C_V},$$

where γ_V is given by (5.31), and we used the reciprocity rule. To establish the last equality we used the result of Problem 5.5.

To evaluate S_{vv} we use again the fundamental equation,

$$S_{vv} = \left(\frac{\partial^2 S}{\partial V^2}\right)_U = \frac{\partial}{\partial V}\left(\frac{P}{T}\right)_U$$

$$= \frac{1}{T}\left(\frac{\partial P}{\partial V}\right)_U - \frac{P}{T^2}\left(\frac{\partial T}{\partial V}\right)_U.$$

Using the result, we find that

$$\left(\frac{\partial P}{\partial V}\right)_U = \frac{-\gamma_V}{C_V}\left(\frac{C_V}{V\beta_P} + T\gamma_V - P\right), \qquad (5.74)$$

with β_P given by (5.29).

On the other hand, from the evaluation of S_{uv} it comes that

$$\left(\frac{\partial T}{\partial V}\right)_U = \frac{T\gamma_V - P}{C_V}. \qquad (5.75)$$

Inserting (5.74) and (5.75) in the equation for S_{vv} one obtains:

$$S_{vv} = -\frac{1}{TVk_T} - \frac{(T\gamma_V - P)^2}{T^2 C_V},$$

with k_T given by (5.33).

Hence, we finally obtain

$$S_{uu} - S_{vv} - (S_{uv})^2 = \frac{1}{T^3 V C_V k_T}.$$

Since $C_V > 0$, the fluting condition for the entropy (5.69) implies that:

$$k_T > 0 \tag{5.76}$$

The inequality expressed by (5.76) determines the criterion for **mechanical stability**: for a thermally stable system to be also mechanically stable, the volume of the system must always decrease in response to a fluctuation that increases the temperature at constant pressure. We can therefore conclude that any fluid system, which is in a state of thermodynamic equilibrium, must exhibit a positive heat capacity at constant V (5.73) and a positive isothermal compressibility at constant T (5.76).

A criterion for **material** or **diffusional stability** can be established if parts A and B are allowed to exchange particles as in Figure 5.4. The interested reader is referred to the textbook of O'Connel and Halle to explore this case.

Also, the reader is likely to anticipate that local stability of the equilibrium state of other thermodynamic potentials will lead to more stability criteria. When considering H and F to derive stability criteria one needs to take into account that one of their natural variables is an intensive state function (P in case of H, and T in the case of F). This means that in equilibrium they take on the same value in each part of the composite system (Figure 5.5). In such cases, local stability requirements involving the potential's derivatives must be obtained indirectly from their Legendre transforms

$$H_{PP} = -1/U_{VV} < 0,$$

and

$$F_{TT} = -1/U_{SS} < 0.$$

The sign of H_{PP} and F_{TT} tells us that H and F are concave functions of their intensive variables. Since, on the other hand, they are convex functions of their extensive variables, the fluting condition (e.g. $H_{SS}H_{PP} - H_{SP}^2 \leq 0$) is true by default. Thus, for these two potentials, the stability requirement reduces to their extensive variables:

$$H_{ss} = \left(\frac{\partial^2 H}{\partial S^2}\right)_{P,N} > 0 \quad \rightarrow \quad \left(\frac{\partial T}{\partial S}\right)_{P,N} > 0, \tag{5.77}$$

$$F_{vv} = \left(\frac{\partial^2 F}{\partial V^2}\right)_{T,N} > 0 \quad \rightarrow \quad \left(\frac{\partial P}{\partial V}\right)_{T,N} < 0. \tag{5.78}$$

The fluting condition for the Gibbs free energy is not trivial and its derivation is outside the scope of the present chapter. To explore these matters in more detail the reader is referred to the textbook of Sekerka.

5.11.2 METASTABILITY OF EQUILIBRIUM STATES

As discussed in the previous section, in a **stable** thermodynamic system the second-order response of the internal energy U (or H, F, G, or $-S$) is positive for all the

differential variations of its natural variables. The most observed thermodynamic
equilibrium states are stable states (Figure 5.6 A). An **unstable** state (Figure 5.6
C), on the other hand, will never be observed because the underlying microscopic
nature of the thermodynamic system ensures that if the system is moved away from
equilibrium by some small fluctuation, there will be another spontaneous fluctuation
that will drive it towards another equilibrium state. For an unstable state, the second-
order response of the internal energy can be made negative for *some* variations of its
natural variables.

Figure 5.6   Schematic representation of the types of equilibrium states found in a simple
mechanical system such as a ball on a gravitational field with potential energy E; x is the
displacement from the equilibrium position. By analogy, the equilibrium states of a thermodynamic system can be classified as stable (A), metastable (B), and unstable (C). In (A) the ball
is in a stable equilibrium because it always returns to the same equilibrium state after any (i.e.
small or large) displacement. In (B) the ball is in a metastable equilibrium and it can return to
it after a small displacement, or move into another equilibrium state after a large displacement.
In (C) the ball is in an unstable equilibrium because the latter will not be maintained after any
displacement. Note that in a thermodynamic system the thermodynamic potential is a function
of at least two variables and the corresponding surface is multidimensional.

Sometimes, it is possible to observe **metastable** states. In mathematical terms a
metastable state is a local minimum (Figure 5.6 B), and differential variations cannot
distinguish it from a stable state, which is a global minimum. In particular, a local
and a global minimum are both characterised by a positive second-order response of
the corresponding potential to differential variations. To determine if an equilibrium
state is metastable, it is necessary to perturb the system with a *finite* (instead of
differential) variation of its natural variables. If the equilibrium state is stable against
all finite variations of the natural variables, then it is a stable equilibrium state (i.e. a
global minimum of U, H, F, G, and $-S$). If *some* finite variations lead to a decrease of
U, H, F, G, or $-S$, then the corresponding equilibrium state classifies as metastable.
The so-called **supercooled** state of water, which is liquid water kept at $T < 0°C$ and
$P = 1$ atm is an example of metastable state. Another example, is that of a substance
kept in a **superheated** state, i.e., in the liquid phase above its boiling point.
    The condition that $d^2U = 0$ (or, alternatively, that $d^2H = 0$, $d^2F = 0$, $d^2G = 0$ or
$d^2S = 0$) provides the **limit of metastability**, and the curve corresponding to this
condition is termed the **spinodal**. When, in the course of a thermodynamic process,
a system moves from an initial equilibrium state that satisfies this condition, it passes
from metastability to instability, and a phase transition is observed. For example, a

tap on a beaker of water in a supercooled state triggers a phase transition whereby the system suddenly and dramatically crystalises. Phase transitions will be discussed in Chapter 7.

5.12   LEARNING OUTCOMES

At the end of this chapter the reader is expected to:

1. Know the extremum principles for the entropy and for the internal energy.
2. Understand that the conditions that are standardly used in the laboratory require the need for other thermodynamic potentials, namely, the free energies.
3. By able to obtain thermodynamic potentials and identify their natural variables by using the Legendre transform.
4. Appreciate the importance of natural variables.
5. Know that Massieu functions are the Legendre transforms of the entropy.
6. Know how to derive Maxwell relations.
7. Appreciate that thermodynamic coefficients are measurable quantities used to quantify the behaviour of a system thermodynamically.
8. Correctly interpret the meaning of chemical potential.
9. Know the fundamental equation of thermodynamics in the internal energy and in the entropy formulation.
10. Be able to formally understand the stability of thermodynamic systems and realise that thermal and mechanical stability of physical systems are a direct consequence of the signs of the second-order derivatives of the internal energy or other thermodynamic potential.
11. Know the difference between stable, unstable, and metastable thermodynamic states.

5.13   WORKED PROBLEMS

PROBLEM 5.1
Consider the coefficient of isobaric expansion β_P, and the coefficient of isothermal compressibility k_T. Starting from $S = S(T, P)$ show that

$$C_P - C_V = \frac{VT\beta_P^2}{k_T}.$$

Solution
We start by recalling from (5.33) and (5.29) that

$$\beta_P = \frac{1}{V}\left(\frac{\partial V}{\partial T}\right)_P,$$

and

$$k_T = -\frac{1}{V}\left(\frac{\partial V}{\partial P}\right)_T.$$

Let us consider $S = S(T, P)$. Since $P = P(T, V)$ we can write $S = S[T, P(T, V)]$.
Thus, be applying the chain rule

$$\left(\frac{\partial S}{\partial T}\right)_V = \left(\frac{\partial S}{\partial T}\right)_P + \left(\frac{\partial S}{\partial P}\right)_T\left(\frac{\partial P}{\partial T}\right)_V.$$

We previously showed that

$$\left(\frac{\partial S}{\partial T}\right)_P = \frac{C_P}{T},$$

and

$$\left(\frac{\partial S}{\partial T}\right)_V = \frac{C_V}{T}.$$

Hence

$$\frac{C_P - C_V}{T} = -\left(\frac{\partial S}{\partial P}\right)_T\left(\frac{\partial P}{\partial T}\right)_V.$$

According to the Maxwell relation for the Gibbs free energy

$$\left(\frac{\partial S}{\partial P}\right)_T = -\left(\frac{\partial V}{\partial T}\right)_P.$$

Therefore, using the definition of β_P, it comes that

$$\frac{C_P - C_V}{T} = V\beta_P\left(\frac{\partial P}{\partial T}\right)_V.$$

Considering $P = P(T, V)$ and using the reciprocity rule one gets

$$\left(\frac{\partial P}{\partial V}\right)_T\left(\frac{\partial V}{\partial T}\right)_P\left(\frac{\partial T}{\partial P}\right)_V = -1. \tag{5.79}$$

By using the reciprocal rule

$$\left(\frac{\partial P}{\partial T}\right)_V = -\left(\frac{\partial P}{\partial V}\right)_T\left(\frac{\partial V}{\partial T}\right)_P.$$

Using again the definition of β_P, and considering the definition of k_T, it comes that

$$\left(\frac{\partial P}{\partial T}\right)_V = \frac{\beta_P}{k_T},$$

Thus, we finally get

$$C_P - C_V = \frac{VT\beta_P^2}{k_T}.$$

PROBLEM 5.2
Show that

$$\left(\frac{\partial T}{\partial P}\right)_H = \frac{T\left(\frac{\partial V}{\partial T}\right)_P - V}{C_P}.$$

Solution

Considering $H = H(T,P)$ and using the reciprocity rule one gets

$$\left(\frac{\partial H}{\partial P}\right)_T \left(\frac{\partial P}{\partial T}\right)_H \left(\frac{\partial T}{\partial H}\right)_P = -1. \tag{5.80}$$

According to the reciprocal rule,

$$\left(\frac{\partial P}{\partial T}\right)_H = \frac{1}{\left(\frac{\partial T}{\partial P}\right)_H},$$

and

$$\left(\frac{\partial T}{\partial H}\right)_P = \frac{1}{\left(\frac{\partial H}{\partial T}\right)_P} = \frac{1}{C_P}.$$

Substituting the two equations above in (5.80)

$$\left(\frac{\partial T}{\partial P}\right)_H = \frac{-\left(\frac{\partial H}{\partial P}\right)_T}{C_P}.$$

Therefore, we need to show that

$$\left(\frac{\partial H}{\partial P}\right)_T = -T\left(\frac{\partial V}{\partial T}\right)_P + V.$$

The differential of $H = H(T,P)$ is

$$dH = \left(\frac{\partial H}{\partial P}\right)_T dP + \left(\frac{\partial H}{\partial T}\right)_P dT. \tag{5.81}$$

On the other hand

$$dH = TdS + VdP. \tag{5.82}$$

Eliminating dH, and solving for dS one obtains

$$dS = \frac{1}{T}\left(\frac{\partial H}{\partial T}\right)_P dT + \frac{1}{T}\left[\left(\frac{\partial H}{\partial P}\right)_T - V\right]dP,$$

with

$$\frac{1}{T}\left[\left(\frac{\partial H}{\partial P}\right)_T - V\right] = \left(\frac{\partial S}{\partial P}\right)_T.$$

Using the Maxwell relation of the Gibbs free energy

$$\left(\frac{\partial S}{\partial P}\right)_T = -\left(\frac{\partial V}{\partial T}\right)_P.$$

Hence,

$$\left(\frac{\partial H}{\partial P}\right)_T = -T\left(\frac{\partial V}{\partial T}\right)_P + V,$$

and finally,

$$\left(\frac{\partial T}{\partial P}\right)_H = \frac{T\left(\frac{\partial V}{\partial T}\right)_P - V}{C_P}.$$

5.14 SUGGESTED PROBLEMS

PROBLEM 5.3
Consider $U = U(S,V)$ and show that

$$\left(\frac{\partial U}{\partial V}\right)_T = T\gamma_V - P.$$

PROBLEM 5.4
Consider $S = S(T,P)$ and show that for a reversible isothermal expansion

$$Q = -VT\beta_P\Delta P.$$

PROBLEM 5.5
Show that

$$\left(\frac{\partial U}{\partial V}\right)_T = T\left(\frac{\partial P}{\partial T}\right)_V - P,$$

and evaluate this expression for the ideal gas. What do you conclude?

PROBLEM 5.6
Consider $U = U(T,V)$ and to show that

$$\left(\frac{\partial T}{\partial V}\right)_U = \frac{P - T\left(\frac{\partial P}{\partial T}\right)_V}{C_V}.$$

PROBLEM 5.7
Use the previous result to show that

$$\left(\frac{\partial C_V}{\partial V}\right)_T = T\left(\frac{\partial^2 P}{\partial T^2}\right)_V.$$

PROBLEM 5.8
Use the identity proven in Problem 5.2 to show that

$$C_P = T\left(\frac{\partial V}{\partial T}\right)_P\left(\frac{\partial P}{\partial T}\right)_S.$$

PROBLEM 5.9
Show that

$$\left(\frac{\partial P}{\partial T}\right)_S = \frac{C_P}{VT\beta_P}.$$

PROBLEM 5.10
Show that

$$\left(\frac{\partial C_P}{\partial P}\right)_T = -T\left(\frac{\partial^2 V}{\partial T^2}\right)_P.$$

PROBLEM 5.11
Show that

$$\frac{C_P}{C_V} = \frac{k_T}{k_S}.$$

PROBLEM 5.12
Show that

$$\frac{\partial}{\partial T}\left(\frac{F}{T}\right) = -\frac{U}{T^2}.$$

PROBLEM 5.13
Use the result obtained in Problem 3.4 to show that

$$F = C_V T - C_V T \ln T - N k_B \ln V - Tc + c',$$

where c' is a constant. Provide an explicit expression for c'.

PROBLEM 5.14
Use the results obtained in the previous problem to show that

$$G = C_P T - C_P T \ln T + N k_B T \ln P - Td + c'$$

where d is a constant. Provide an explicit expression for d.

PROBLEM 5.15
Consider an hypothetical fluid for which

$$U(S,V) = a\frac{S^2}{V},$$

where a is a constant. Determine the equations of state that assume the form
$T = T(S,V)$ and $P = P(S,V)$.

PROBLEM 5.16
Show that C_P is always positive.

PROBLEM 5.17
Show that the isothermal compressibility k_T is always positive by using the
stability criteria for the Helmholtz free energy.

REFERENCES

1. Callen, H. B. (1960). Thermodynamics. Wiley.

2. Dill, K. A. & Bromberg S. (2002). Molecular Driving Forces: Statistical Thermodynamics in Biology and Chemistry. Taylor & Francis.

3. Kondepudi, D. (2008). Introduction to Modern Thermodynamics: From Heat Engines to Dissipative Structures. Wiley.

4. O' Connel, J. P. & Haile, J. M. (2005). Thermodynamics: Fundamentals and Applications. Cambridge University Press.

5. Reiss, H. (2012). Methods of Thermodynamics. Dover.

6. Scott Shell, M. (2015). Thermodynamics and Statistical Mechanics: An Integrated Approach. Cambridge University Press.

7. Sekerka, R. F. (2015). Thermal Physics: Thermodynamics and Statistical Mechanics for Physicists and Engineers. Elsevier.

8. Swendsen, R. H. (2012). An Introduction to Statistical Mechanics and Thermodynamics. Oxford University Press.

6 Thermodynamics of Extensive Systems

This chapter is dedicated to derive general thermodynamic relations. Special focus is placed on understanding the consequences of extensivity for the formal structure of thermodynamics. The importance of molar quantities and partial molar quantities is also discussed.

INTRODUCTION

In the first chapter of this textbook we discussed the difference between extensive and intensive property. Here, we will explore the consequences of extensivity for thermodynamics. However, before doing so, we should analyse the requirement that a system must be homogeneous for extensivity to hold. Let us then consider a gas that is enclosed in a cylindrical vessel tapped by a piston. This time, however, the gas is not ideal and apart from interacting with each other, the composing particles also interact with the internal surface of the vessel. If the interaction with the surface is strong enough, some particles will become adsorbed at the surface, and the corresponding intermolecular interactions will contribute to the systems's total internal energy. Does extensivity hold in this scenario? In other words, can we say that for such a system the internal energy, the volume, the number of particles, etc. scale linearly with the systems' size? The answer is *no*. Indeed, if the size of the cylinder changes, the surface-to-volume ratio changes, and the fraction of adsorbed particles will also change. Therefore, the properties of the system will depend on the surface-to-volume ratio, and will not be extensive. It is interesting to note that many thermodynamics textbooks assume tacitly that systems are homogeneous, but in most real cases (such as that of the simple example described above) they are not. Extensivity will not hold whenever boundary effects exist. On the other hand, if one is interested in the thermodynamic behaviour of the bulk of the system without taking into account what happens at surfaces and interfaces, then it is reasonable to assume that the system, being homogeneous, will satisfy the postulate of extensivity. In what follows we will be assuming that extensivity holds and determine the consequences of extensivity for thermodynamics.

6.1 THE EULER EQUATION

In thermodynamics, the Euler equation is a consequence of Euler's theorem, which is a general statement about homogeneous functions of degree k. In particular, Euler's equation follows from the application of Euler's theorem to homogeneous functions of degree one, which are extensive functions, as discussed in Chapter 1.

DOI: 10.1201/9781003091929-6

EULER THEOREM

If $X(x,y,z)$ is an homogeneous function of degree k,

$$x\left(\frac{\partial X}{\partial x}\right)_{y,z} + y\left(\frac{\partial X}{\partial y}\right)_{x,z} + z\left(\frac{\partial X}{\partial z}\right)_{x,y} = kX(x,y,z). \tag{6.1}$$

The proof of the Euler theorem is straightforward. Consider an homogeneous function $X(x,y,z)$ of degree k. Then,

$$X(\lambda x, \lambda y, \lambda z) = \lambda^k X(x,y,z).$$

Taking the partial derivative with respect to λ one gets

$$\frac{\partial X(\lambda x,\lambda y,\lambda z)}{\partial(\lambda x)}\frac{\partial(\lambda x)}{\partial\lambda} + \frac{\partial X(\lambda x,\lambda y,\lambda z)}{\partial(\lambda y)}\frac{\partial(\lambda y)}{\partial\lambda} + \frac{\partial X(\lambda x,\lambda y,\lambda z)}{\partial(\lambda z)}\frac{\partial(\lambda z)}{\partial\lambda} = k\lambda^{(k-1)}X(x,y,z).$$

Noticing that $\partial(\lambda x)/\partial\lambda = x$, $\partial(\lambda y)/\partial\lambda = y$, and $\partial(\lambda z)/\partial\lambda = z$, and taking $\lambda = 1$ the result follows.

In the demonstration above we used a function of three variables, but the theorem holds for any number of independent variables. It is also important to remark that if X depends on additional variables such that $X(\lambda x, \lambda y, \lambda z, a, b) = \lambda^k X(x,y,z,a,b)$, equation (6.1) still holds, without the corresponding terms for a and b on the left-hand side.

Let us now apply the Euler theorem to the internal energy U, which is an extensive function, i.e., an homogeneous function of degree one ($k = 1$) in the extensive variables S, V, and N. Let us consider the most general case in which the system is composed of N particles of m different types, such that

$$N = N_1 + N_2 + ... + N_m.$$

In that case U is an extensive function of S, V, and $N_1, N_2, ..., N_m$:

$$U = U(S,V,N_1,N_2,...,N_m).$$

From equation (6.1) with $k = 1$ it comes that

$$S\left(\frac{\partial U}{\partial S}\right)_{V,N} + V\left(\frac{\partial U}{\partial V}\right)_{S,N} + \sum_{i=1}^{m} N_i\left(\frac{\partial U}{\partial N_i}\right)_{S,V,\{N_{j\neq i}\}} = U.$$

Using the thermodynamic definitions of temperature, pressure, and chemical potential one obtains the **Euler equation** for the internal energy:

$$TS - PV + \sum_{i=1}^{m} N_i\mu_i = U \tag{6.2}$$

Substituting the Euler equation in the definition of the Gibbs free energy (5.19) one obtains

$$G = \sum_{i=1}^{m} N_i \mu_i. \tag{6.3}$$

Note that one could have derived (6.3) directly from the application of Euler's theorem to $G = G(T, P, N_1, N_2, ..., N_m)$. This is left as an exercise.

In the case of a one-component system equation (6.3) simplifies to

$$G = \mu N, \tag{6.4}$$

and we can interpret the chemical potential as being the Gibbs free energy per particle. Dividing equation (6.3) by N, one obtains the **molar Gibbs free energy**

$$g \equiv \frac{G}{N} = \sum_{i=1}^{m} x_i \mu_i, \tag{6.5}$$

with

$$x_i \equiv \frac{N_i}{N}, \tag{6.6}$$

being the so-called **molar fractions**.

Since

$$\sum_{i=1}^{m} x_i = 1, \tag{6.7}$$

there are only $(m-1)$ independent molar fractions.

An equation similar to (6.2) can be derived for the the entropy, which is an extensive variable of U, V, N_1, N_2, ..., N_m:

$$S = S(U, V, N_1, N_2, ..., N_m).$$

In this case the application of Euler theorem leads to the **Euler equation** for the entropy:

$$S = \left(\frac{1}{T}\right) U + \left(\frac{P}{T}\right) V - \left(\frac{1}{T}\right) \sum_{i=1}^{m} N_i \mu_i \tag{6.8}$$

6.2   THE GIBBS-DUHEM EQUATION AND THERMODYNAMIC DEGREES OF FREEDOM

The Gibbs-Duhem equation is easily derived from the Euler equation. We start by taking the differential of equation (6.2):

$$dU = TdS + SdT - PdV - VdP + \sum_{i=1}^{k} dN_i \mu_i + \sum_{i=1}^{m} N_i d\mu_i.$$

By subtracting from the equation above the fundamental constraint as given by equation (5.44), one obtains the **Gibbs-Duhem equation** for the internal energy of a multicomponent system:

$$0 = SdT - VdP + \sum_{i=1}^{m} N_i d\mu_i \qquad (6.9)$$

Equation (6.9) shows that changes in the intensive variables T and P, and changes in the m chemical potentials $\mu_1, \mu_2, ..., \mu_m$ cannot all be independent. In other words, T, P, and $\mu_1, \mu_2, ..., \mu_m$ are not all independent variables.

Consider the case in which the system is formed by one particle type only ($m = 1$) and rewrite the equation (6.9) as

$$d\mu = -sdT + vdP, \qquad (6.10)$$

where

$$s = \frac{S}{N}, \qquad (6.11)$$

is the **molar entropy**, and

$$v = \frac{V}{N} \qquad (6.12)$$

is the **molar volume**.

Other molar quantities such as the **molar internal energy** or **the molar entropy**, are similarly defined:

$$u = \frac{U}{N}, \qquad (6.13)$$

and

$$s = \frac{S}{N}. \qquad (6.14)$$

Molar quantities are sometimes termed **specific quantities** (e.g. specific volume).

Equation (6.10) expresses the change in the chemical potential as a function of a change in T and in P,

$$\mu = \mu(T, P).$$

Thus, for a one-component system, there are only two independent intensive variables.

For a multicomponent system with m types of particles, there will be $(m + 1)$ independent intensive variables he number of independent intensive variables that are needed to specify the thermodynamic state of a system is the number of **thermodynamic degrees of freedom**.

An equivalent version of the Gibbs-Duhem equation can be obtained for the entropy. In this case one takes the differential of equation (6.8) and subtracts the differential of S as given by equation (5.46) to obtain

$$U d\left(\frac{1}{T}\right) + V d\left(\frac{P}{T}\right) - \sum_{i=1}^{m} N_i d\left(\frac{\mu_i}{T}\right) = 0 \qquad (6.15)$$

Think about it...
Derive the Gibbs-Duhem equation by considering the enthalpy of a multicomponent system.

Answer

Since $H = H(S, P, N_1, N_2, ..., N_m)$, with P being intensive, the application of Euler's theorem to H leads to

$$TS + \sum_{i=1}^{m} \mu_i N_i = H.$$

The differential of H is thus

$$dH = T dS + S dT + \sum_{i=1}^{m} \mu_i dN_i + \sum_{i=1}^{m} d\mu_i N_i.$$

On the other hand

$$dH = T dS + V dP + \sum_{i=1}^{m} \mu_i dN_i.$$

By subtracting the last two equations one gets

$$0 = S dT - V dP + \sum_{i=1}^{m} N_i d\mu_i,$$

which is the Gibbs-Duhem equation.

3   APPLYING THE GIBBS-DUHEM EQUATION

 what follows we apply the Gibbs-Duhem equation to determine the chemical potential and the entropy equation of state of the ideal gas. In Chapter 7, we will illustrate the importance of the Gibbs-Duhem equation while studying phase transitions.

We start by considering the Gibbs-Duhem equation for the entropy of a one-component system

$$d\left(\frac{\mu}{T}\right) = u d\left(\frac{1}{T}\right) + v d\left(\frac{P}{T}\right)$$

with u and v being molar quantities.

According to the ideal gas pressure equation of state

$$PV = Nk_BT \Leftrightarrow \frac{P}{T} = \frac{k_B}{v},$$

(6.16)

with v being given by equation (6.12).

Using the equation of state for the internal energy

$$U = \frac{3}{2}Nk_BT \Leftrightarrow \frac{1}{T} = \frac{3}{2}\frac{k_B}{u},$$

(6.17)

with u given by (6.13). Therefore,

$$d\left(\frac{P}{T}\right) = -\frac{k_B}{v^2}dv,$$

and

$$d\left(\frac{1}{T}\right) = -\frac{3}{2}k_B\frac{du}{u^2}.$$

The Gibbs-Duhem equation for the ideal gas is then

$$d\left(\frac{\mu}{T}\right) = -\frac{3}{2}k_B\frac{du}{u} - k_B\frac{dv}{v}.$$

By integrating the equation above one obtains **the equation of state for the chemical potential** of the ideal gas

$$\frac{\mu}{T} = \frac{\mu_0}{T} - \frac{3}{2}k_B\ln\left(\frac{u}{u_0}\right) - k_B\ln\left(\frac{v}{v_0}\right),$$

(6.18)

where (μ_0, u_0, v_0) specify a reference state.

For a single component system the Euler equation for the entropy (6.8) can be written as

$$s = -\frac{\mu}{T} + \left(\frac{u}{T}\right) + \left(\frac{Pv}{T}\right),$$

(6.19)

with s being the molar entropy (6.14).

Using equations (6.16) and (6.17), equation (6.19) simplifies to

$$s = -\frac{\mu}{T} + \frac{5}{2}k_B,$$

with μ/T being given by equation (6.18).

Therefore, **the entropy equation of state for the ideal gas is**

$$s = s_0 + \frac{3}{2}k_B\ln\left(\frac{u}{u_0}\right) + k_B\ln\left(\frac{v}{v_0}\right),$$

(6.20)

with

$$s_0 = \frac{5}{2}k_B - \frac{\mu_0}{T}.$$

Think about it...

From a formal point of view what is the difference between the ideal gas equations $PV = Nk_BT$, $U = 3/2Nk_BT$, and equation (6.20) ?

Answer

Although all of them are equations of state, only the one for the entropy is a fundamental equation of state, expressing s as function of its natural variables u and v.

4   MOLAR QUANTITIES

olar quantities are intensive properties because they represent the ratio between ~o extensive quantities. What is the mathematical consequence of working with ~olar quantities? In order to answer this question let us consider the example of the ~ternal energy. Since U is extensive in all its natural variables, it is true that

$$U(\lambda S, \lambda V, \lambda N_1, \lambda N_2, ..., \lambda N_m) = \lambda U(S, V, N_1, N_2, ..., N_m).$$

n the other hand, taking into account that λ is arbitrary, we can take $\lambda = 1/N$ and ~tain

$$U\left(\frac{S}{N}, \frac{V}{N}, \frac{N_1}{N}, \frac{N_2}{N}, ..., \frac{N_m}{N}\right) = \frac{U(S, V, N_1, N_2, ..., N_m)}{N}.$$

iven the definition of molar entropy (6.14), molar volume (6.12), molar fraction ~.6), and molar internal energy (6.13), the equality above can be rewritten as

$$U(s, v, x_1, x_2, ..., x_m) = u. \tag{6.21}$$

~quation (6.21) can expressed in terms of the $(m-1)$ independent molar fractions ~.7) as

$$u(s, v, x_1, x_2, ..., x_{m-1}) = u. \tag{6.22}$$

~nus, the mathematical consequence of using molar quantities is that:

When working with molar quantities the number of independent variables is re-duced by one

---

~his can be seen by considering the differential of u:

$$du = Tds - Pdv + \sum_{i=1}^{m-1} (\mu_i - \mu_m)dx_i, \tag{6.23}$$

~hose derivation is presented in worked Problem 6.2.

For a one-component system (for which $dx_i = 0$) equation above reduces to

$$du = Tds - Pdv,$$ (6.24)

with $u = u(s,v)$.

The approach described above can be used for other molar quantities, and equations equivalent to (6.23) and (6.24) can be obtained for s, h, f, and g. We leave this as an exercise for the reader.

6.5 PARTIAL MOLAR QUANTITIES

When studying multicomponent systems (such as liquid mixtures and metal alloys), each particle species can have molar properties that are different from the molar properties of the other species. This is a consequence of the fact that the number of particles $N_1, N_2, ..., N_m$ can be specified independently and different particles can be present in different amounts. Thus, a single molar property such as v is no longer enough to specify V. In this case, in order to determine how a certain extensive thermodynamic property depends on the different components, it is useful to consider the so-called partial molar properties.

Consider a multicomponent system, and a generic thermodynamic extensive property Y, that is expressed as a function of P, T and $N_1, N_2, ..., N_m$:

$$Y = Y(P,T,N_1,N_2,...,N_m).$$

Since Y is an extensive function in $N_1, N_2, ..., N_m$, it is true that:

$$Y(T,P,\lambda N_1,\lambda N_2,...,\lambda N_m) = \lambda Y(T,P,N_1,N_2,...,N_m).$$

Applying Euler's theorem we obtain

$$Y(T,P,N_1,N_2,...,N_m) = \sum_{i=1}^{m} \bar{y}_i N_i,$$ (6.25)

where $\bar{y}_i$ are intensive properties called **partial molar quantities**:

$$\bar{y}_i = \left(\frac{\partial Y}{\partial N_i}\right)_{P,T,\{N_{j \neq i}\}}$$ (6.26)

Thus, $\bar{y}_i$ represents the change in Y that is observed upon adding a small amount of species i to the system while keeping P, T, and the amounts of all the other species constant. Partial molar quantities depend on temperature, pressure, and composition. They are intensive state functions.

Think about it...
Is the chemical potential

$$\mu_i = \left(\frac{\partial U}{\partial N_i}\right)_{S,V,\{N_{j\neq i}\}}$$

a partial molar quantity?

Answer
No, because in the definition of partial molar quantities the constraints are T, P, and all the other particle numbers except N_i, and in the definition above the constraints are S, V, and all the other particle numbers except N_i.

The chemical potential is a partial molar property only when it is defined from the Gibbs free energy:

$$\mu_i = \left(\frac{\partial G}{\partial N_i}\right)_{T,P,\{N_{j\neq i}\}}.$$

Note that since P and T are held constant in the definition of partial molar quan-
ies one can differentiate equations (5.2), (5.13), and (5.20) to obtain $\bar{h}_i = \bar{u}_i + P\bar{v}_i$,
$= \bar{u}_i - T\bar{s}_i$, and $\bar{g}_i = \bar{u}_i - T\bar{s}_i + P\bar{v}_i$.

To analyse the consequences of extensivity in N, let us focus in the case of the
lume. According to equation (6.25)

$$V(T,P,N_1,N_2,...,N_m) = \sum_{i=1}^{m} \bar{v}_i N_i, \tag{6.27}$$

ith

$$\bar{v}_i = \left(\frac{\partial V}{\partial N_i}\right)_{P,T,\{N_{j\neq i}\}}.$$

or many mixtures the volume is independent of the composition of the mixture, but
r several others it is not. An example, is a mixture of water and alcohol, where for
nall amounts of the latter the volume of the mixture is actually smaller than the sum
the volumes of water and alcohol.

By computing the differential of V from equation (6.27) one obtains

$$dV = \sum_{i=1}^{m} \bar{v}_i dN_i + \sum_{i=1}^{m} d\bar{v}_i N_i. \tag{6.28}$$

n the other hand, since $V = V(T,P,N_1,N_2,...,N_m)$, the differential of V is also given

$$dV = \left(\frac{\partial V}{\partial T}\right)_{P,\{N_i\}} dT + \left(\frac{\partial V}{\partial P}\right)_{P,\{N_i\}} dP + \sum_{i=1}^{m} \left(\frac{\partial V}{\partial N_i}\right)_{T,P,\{N_{j\neq i}\}} dN_i.$$

y using equations (5.29), (5.33), and (6.26) the previous equation can be rewritten

$$dV = V\beta_P dT - Vk_T dP + \sum_{i=1}^{m} \bar{v}_i dN_i, \tag{6.29}$$

which simplifies to

$$dV_{(T,P)} = \sum_{i=1}^{m} \bar{v}_i dN_i,$$ (6.30)

at constant temperature and pressure, as indicated by the subscripts in dV.

For equation (6.30) to be consistent with (6.28) it is necessary that at constant temperature and pressure

$$\sum_{i=1}^{m} d\bar{v}_i N_i = 0.$$ (6.31)

It is possible to obtain equations similar to (6.27) and (6.31) for other state functions that are extensive in the number of particles. In the case of the Gibbs free energy, for example, the equivalent equations are

$$G(T,P,N_1,N_2,...,N_m) = \sum_{i=1}^{m} \bar{g}_i N_i = \sum_{i=1}^{m} \mu_i N_i$$ (6.32)

and

$$\sum_{i=1}^{m} d\bar{g}_i N_i = \sum_{i=1}^{m} d\mu_i N_i = 0.$$ (6.33)

Equations (6.31) and (6.33) show that at constant T and P, the partial molar volumes $d\bar{v}_i$, and the chemical potentials μ_i are not independent of each other. The reader may recall that the non-interdependence of the chemical potentials is expressed by the Gibbs-Duhem equation (6.9) when the temperature and pressure are both kept constant.

6.6   LEARNING OUTCOMES

At the end of this chapter the reader is expected to:

1. Use Euler theorem to derive the Euler equation.
2. Know how to derive the Gibbs-Duhem equation and understand the concept of thermodynamic degrees of freedom.
3. Understand the importance of molar quantities and distinguish them from partial molar quantities.

6.7   WORKED PROBLEMS

PROBLEM 6.1
Derive an expression for dg at constant T and P.

Solution

The differential of equation (6.5) is

$$dg = \sum_{i=1}^{m} x_i d\mu_i + \sum_{i=1}^{m} dx_i \mu_i.$$

Dividing the Gibbs-Duhem equation (6.9) by N, and rearranging one obtains

$$\sum_{i=1}^{m} x_i d\mu_i = v dP - s dT.$$

Substituting the equation above in the equation for dg,

$$dg = v dP - s dT + \sum_{i=1}^{m} \mu_i dx_i.$$

Finally, if P and T are fixed

$$dg_{(T,P)} = \sum_{i=1}^{m} \mu_i dx_i.$$

PROBLEM 6.2
Derive equation (6.23).

Solution
The differential of u is given by

$$du = -\frac{U}{N^2} dN + \frac{dU}{N}$$

Using Euler's equation (6.2) and the fundamental constraint (5.44) the equation above can be written as

$$du = -\frac{TS - PV + \sum_{i=1}^{m} \mu_i N_i}{N^2} dN + \frac{T dS - P dV + \sum_{i=1}^{m} \mu_i dN_i}{N}.$$

Considering that

$$ds = \frac{dS}{N} - \frac{S}{N^2} dN \Leftrightarrow \frac{dS}{N} = ds + \frac{S}{N^2} dN,$$

$$dv = \frac{dV}{N} - \frac{V}{N^2} dN \Leftrightarrow \frac{dV}{N} = dv + \frac{V}{N^2} dN,$$

and

$$dx_i = \frac{dN_i}{N} - \frac{N_i}{N^2} dN \Leftrightarrow \frac{dN_i}{N} = dx_i + \frac{N_i}{N^2} dN,$$

and substituting the corresponding terms for dS/N, dV/N, and dN_i/N on the right-hand side of the equation for du, the latter can be rewritten as

$$du = T ds - P dv + \sum_{i=1}^{m} \mu_i dx_i.$$

Taking into account that there are only $(m-1)$ independent molar fractions $\{x_i\}$, one finally obtains equation (6.23):

$$du = T ds - P dv + \sum_{i=1}^{m-1} (\mu_i - \mu_m) dx_i.$$

6.8   SUGGESTED PROBLEMS

PROBLEM 6.3

Show that the function $f(x,y,z) = (x^2 + y^2 + z^4/x^2 + y^2)$ is a homogeneous function of degree 2 in x, y, and z.

PROBLEM 6.4

Show that the area of a sphere, $A = 4\pi R^2$, can be considered a homogeneous function of degree $2/3$, such that $A(\lambda V) = \lambda^{2/3} A(V)$.

PROBLEM 6.5

Apply Euler's theorem to $H = H(S, P, N_1, N_2, ..., N_m)$. What do you conclude?

PROBLEM 6.6

Derive the fundamental equation for the molar entropy, ds, and for the molar enthalpy, dh of a multicomponent system of m different species.

PROBLEM 6.7

Use the Gibbs-Duhem equation for a multicomponent system of m different species to show that

$$\sum_{i=1}^{m} \left(\frac{\partial \mu_k}{\partial N_i} \right)_{P,T} N_i = 0.$$

REFERENCES

1. Kondepudi, D. (2008). Introduction to Modern Thermodynamics: From Heat Engines to Dissipative Structures. Wiley.

2. O' Connel J. P., Haile, J. M. (2005) Thermodynamics: Fundamentals and Applications. Cambridge University Press.

3. Sekerka, R. F. (2015) Thermal Physics: Thermodynamics and Statistical Mechanics for Physicists and Engineers. Elsevier.

4. Swendsen, R. H. (2012) An Introduction to Statistical Mechanics and Thermodynamics. Oxford University Press.

Section III

Applications

7 Phase Transitions

This chapter provides a brief introduction to the study of phase transitions in the context of thermodynamics. Focus is placed on the van der Waals gas and on the analysis of a very particular point of the phase diagram called critical point. In doing so, it highlights the applicability of the concepts and tools of thermodynamics presented in the first and second parts of this book.

7.1 INTRODUCTION

Throughout this book we have been often invoking the ideal gas model. Its simplicity allows one to perform simple analytical calculations that are extremely useful to apply and learn the theory of thermodynamics. Furthermore, the physical behaviour of a diluted gas can be easily interpreted on the basis of the underlying microscopic model. However, we must insist that thermodynamics is a remarkaby general subject with an outstanding broad scope of applicability. One example is the study of phase transitions. Simply put, a **phase transition** is the process according to which a system changes its physical state as a result of changing some thermodynamic parameter in the absence of chemical reactions (e.g. water vapour condensing into liquid water when the temperature is lowered at standard atmospheric pressure).

As we have seen in Chapter 5, at constant temperature and pressure, a closed system evolves towards an equilibrium state where $G = H - TS$ is a minimum. When the temperature is high, the entropic term $(-TS)$ prevails and the equilibrium state is the one dominated by entropy (e.g. the gas phase); at lower temperature the enthalpic term dominates and the equilibrium state is the one that minimises the enthalpy (e.g. the liquid phase).

Phase transitions require the existence of intermolecular interactions between the system's particles. The strength of intermolecular interactions relative to **thermal energy** $(k_B T)$ becomes progressively larger as one moves from the gas to the liquid phase, and from the latter to the solid phase. Indeed, in liquids, the relative strength of attractive intermolecular interactions is large enough to keep the particles much closer than in a gas, and almost as close as in a solid. However, the particles exhibit no regular arrangement, while in a solid they are locked into the positions of a crystalline lattice, vibrating around their equilibrium positions.

The dutch scientist Johannes van der Waals (1837–1923) was the first to develop a model of a fluid system that takes into account the effects of intermolecular interactions in the system's behaviour. As we shall see, despite its simplicity, the van der Waals model is – up to a certain extent – able to predict phase transitions such as the one converting a gas into a liquid, and it is also able to predict the existence of the **critical point**. The latter is a peculiar point of the phase diagram dominated by **fluctuations** that occur at every scale (e.g. density fluctuations in the case of the gas-liquid phase transition), including the macroscopic scale of the system. To study this

DOI: 10.1201/9781003091929-7

unique point of the phase diagram a new physical theory is required that correctly captures and quantitatively predicts the system's behaviour.

7.2 PHASE

A **phase** is a part of a system, physically distinct, macroscopically homogeneous, and containing one or several components (i.e. particles of different types). It is physically separable from the rest of the system.

Consider the case of a thermodynamic system that is exclusively composed of water molecules. We know from our daily experience that at standard atmospheric pressure water may be in the solid, liquid, or gas phase, depending on the temperature. We recognise ice, liquid water, and gaseous water as three different phases of the same substance. If a system consists of liquid water and ice cubes in a glass jar, we identify the ice as being one phase and liquid water as being another phase. Indeed, as we will see briefly, under certain conditions different phases may coexist in contact with each other in thermodynamic equilibrium.

Other, perhaps less familiar examples of thermodynamic phases include plasmas, superfluid helium, and superconductivity. A plasma is a gas composed of charged particles (electrons and ions) obtained by superheating a gas between several thousand and several million kelvin, i.e., in extreme conditions which are not naturally found on Earth. Superfluid helium is characterised by the absence of viscosity when the temperature approaches absolute zero, and the superconductivity of certain materials (e.g. certain metals and alloys), is characterised by the absence of electrical resistance.

Think about it...
Consider a system formed by water and oil at room temperature. How many phases are there in the system? And in a mixture of alcohol and water?

Answer

In principle, at room temperature the oil will not mix with water and it will be possible to identify two phases. On the other hand, water and alcohol are fully miscible and the resulting mixture will have only one phase.

7.3 THE GIBBS PHASE RULE

Gibbs derived an important relationship among the number of components m, the number of phases, φ, and the number of intensive variables, f, that must be specified in order to characterise the equilibrium state of a non-reacting multicomponent system. The latter is a system a system in which the passage of a component from one phase to another does not involve any chemical reaction. In particular, the **Gibbs phase rule**, or simply **phase rule**, determines the number of independent intensive variables (or thermodynamic degrees of freedom) as a function of φ and m according to

$$f = m + 2 - \varphi \tag{7.1}$$

The derivation of the phase rule following Gibbs is straightforward. Let us consider a closed system in equilibrium that consists of parts α, β, ..., φ. Each part has its own temperature and pressure, and contains particles of species $i = 1, 2, ..., m$ (Figure 7.1). Let us recall the equilibrium conditions for a thermodynamic system composed

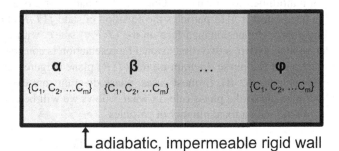

Figure 7.1 A thermodynamic system that is formed by several phases α, β, ..., φ. Each phase has m components, $C_1, C_2, ..., C_m$ that can be exchanged between the different phases in the absence of chemical reactions.

of different parts from equations (5.60)–(5.62). Applying those conditions to our multiphase system yields:

$$T_\alpha = T_\beta = ... = T_\varphi = T,$$

$$P_\alpha = P_\beta = ... = P_\varphi = P,$$

and

$$\mu_{1,\alpha} = \mu_{1,\beta} = ... = \mu_{1,\varphi},$$

$$\mu_{2,\alpha} = \mu_{2,\beta} = ... = \mu_{2,\varphi},$$

$$...$$

$$\mu_{m,\alpha} = \mu_{m,\beta} = ... = \mu_{m,\varphi}.$$

Therefore, the multiphase system is characterised by $(m + 2)$ variables.

However, the $(m + 2)$ variables are not all independent since there is one Gibbs-Duhem equation (6.9) per phase relating the intensive variables T and P with the chemical potentials of the different species,

$$0 = SdT - VdP + \sum_{i=1}^{m} N_i d\mu_i.$$

Indeed, to get the number of independent variables of the multiphase system one must subtract from $(m+2)$ the number of Gibbs-Duhem equations, which is φ. The system is then characterised by $f = (m+2) - \varphi$ independent variables. In the case of a single-component system, the phase rule reduces to

$$f = 3 - \varphi.$$

7.4 PHASE DIAGRAMS

A **phase diagram** is a graphical representation of the equation of state. Consider, for example, the general case of a multicomponent fluid system. In this case the equation of state is $f(T,P,V,N_1,N_2,...,N_m) = 0$ and it is impossible to graphically represent the corresponding multidimensional surface. One the other hand, for a one-component system with a fixed number N of particles the equation of state $f(T,P,v) = 0$ can be represented by a three-dimensional surface on the (T,P,v) space, with $v = V/N$ being the molar volume. Often, a two-dimensional representation is considered, which shows a projection of the phase diagram on the (T,P) plane (Figure 7.2 A), or on the (v,P) plane (Figure 7.2 B). In most cases, phase diagrams are experimentally determined with the aid of the phase rule. In what follows we will be considering a one-component fluid with a fixed number of particles.

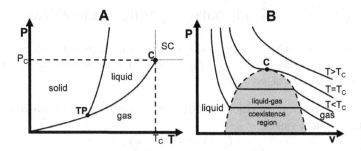

Figure 7.2 Phase diagram of a single-component fluid projected on the (T,P) state plane (A) and on the (v,P) state plane (B). In the (T,P) projection there are three coexistence lines (solid-liquid, solid-gas, and liquid-gas) that intersect at the triple point (TP). The liquid-gas coexistence line terminates at the so-called critical point (C). Above T_C and P_C the system exists as a supercritical (SC) fluid, rather than a supercritical gas or liquid. The (v,P) projection contains the isothermal curves. Below the critical temperature T_C, the isotherms exhibit an horizontal section which is a zone of phase coexistence. Above T_C the isotherms resemble those of the ideal gas.

In the (T,P) projection of the phase diagram there are three areas that correspond to the solid, liquid, and gas phases. According to the phase rule, the number of degrees of freedom of a one-component system $(m = 1)$ with one phase is $f = 2$, which means that the temperature and pressure can both be independent.

The so-called coexistence lines in the (T,P) projection separate the areas corresponding to the single phases and represent the coexistence of two-phases: solid and liquid (**melting line**), solid and gas (**sublimation line**), and liquid and gas (**boiling line**). By using the phase rule, the number of degrees of freedom of a one-component system with two phases is $f = 1$. In this case the temperature must be a function of pressure $T = T(P)$, or vice-versa. The melting line, sublimation line, and boiling line thus respectively provide the melting temperature, the sublimation temperature and the boiling temperature at any pressure. The liquid-gas coexistence line is also termed vapour-pressure line because it provides the maximum pressure the system can stand as a gas for a given temperature. The three coexistence lines intersect at a the **triple point** in which the three phases coexist in equilibrium. Since in this case $f = 0$ the TP stays determined by a unique T_{TP} and P_{TP}. The crossing of a coexistence line while moving from one area (i.e. from one phase) to another corresponds to the occurrence of a phase transition (e.g. sublimation and deposition; vaporisation and condensation; melting and freezing).

The boiling line, and only it, terminates at a particular point termed **critical point**, which is characterised by a critical temperature T_C and a critical pressure P_C. For $P > P_C$ and $T > T_C$ the system is in the so-called **supercritical state**. In this case it is impossible to distinguish whether the system is a liquid or a gas. As a result, supercritical fluids do not have a definite phase.

Note that beyond the critical point it is possible to continuously transform a liquid into a gas without crossing the liquid-gas coexistence line. We will discuss the critical point of the phase diagram in more detail later on.

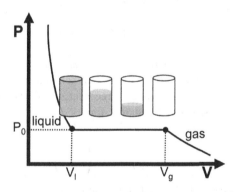

Figure 7.3   Shape of a liquid-gas isotherm below the critical temperature, which highlights the region corresponding to the liquid phase, the region where the liquid and gas coexist, and the region corresponding to the gas phase. P_0 is designated by vapour pressure.

7.5   PHASE TRANSITIONS

In order have a better feeling of what a phase transition is, we will now look into the (v, P) projection of the phase diagram, and, in particular, we analyse the **liquid-gas isotherm** (Figure 7.3). A similar analysis leads to similar conclusions for the liquid-solid and solid-gas isotherms.

In isotherms the temperature is constant. The liquid-gas isotherm contains three different parts. The part where the absolute value of $(\partial P/\partial v)_T$ is large, that is, the pressure increases sharply with a reducing volume, corresponds to the relatively incompressible liquid phase. The part where the pressure decreases smoothly with an increasing volume, that is, the absolute value of $(\partial P/\partial v)_T$ is small, corresponds to the highly compressible gas phase. Finally, the intermediate region at constant **vapour pressure** P_0 corresponds to the transition region in which the liquid (l) and gas (g) coexist in equilibrium. Note that when projected on the (T,P) state plane the coexistence region of the isotherms appears as the lines separating the different phases, i.e., the coexistence lines. In the coexistence region some particles are in the liquid phase, while others are in the gas phase such that $N_l + N_g = N$. In point (P_0, v_l) of Figure 7.3 the particles are all in the liquid phase with density $\rho_l = N/V_l$, while in point (P_0, v_g) the particles are all in the gas phase with density $\rho_g = N/V_g$. Since the two phases have different densities the total volume changes abruptly at constant pressure (or as a result of an infinitesimal change in pressure) form v_l to $v_g \gg v_l$ or vice-versa, as the relative amount of particles in the liquid and gas phases changes.

Since the (molar) volume v is thermodynamically defined by

$$v = \left(\frac{\partial g}{\partial P} \right)_T,$$

a discontinuous change in v means that the first derivative of the (molar) Gibbs free energy with respect to pressure is not a regular function. The same observation holds for the first derivative of the (molar) Gibbs free energy with respect to temperature,

$$s = -\left(\frac{\partial g}{\partial T} \right)_P,$$

as a result of a discontinuous change in (molar) entropy s when the system moves from the liquid to the gas phase, or vice-versa. When at least one of the first derivatives of the molar (Gibbs or Helmholtz) free energy are discontinuous, the phase transition is termed a **first-order** phase transition.

Since the isothermal compressibility is defined as

$$k_T v = -\left(\frac{\partial v}{\partial P} \right)_T = -\left(\frac{\partial^2 g}{\partial P^2} \right)_T,$$

and the (molar) heat capacity at constant pressure is defined as

$$\frac{c_P}{T} = \left(\frac{\partial s}{\partial T} \right)_P = -\left(\frac{\partial^2 g}{\partial T^2} \right)_P,$$

a discontinuity in v and s must cause a **singularity** (i.e. a point at which it becomes infinite) in k_T and c_P, respectively. Phase transitions are thus described theoretically by the appearance of singularities (non-analyticities) in functions representing thermodynamic quantities that are generally used to characterise a thermodynamic system. A phase, on the other hand, is described by the Gibbs free energy, or by the Helmholtz free energy, which are analytical functions of its natural variables.

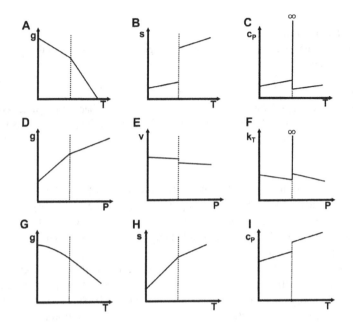

Figure 7.4 Types of phase transitions. Schematic representation of the temperature dependence of the molar Gibbs free energy, molar entropy, and molar heat capacity at constant
pressure in a first-order phase transition (A–C). Pressure dependence of the molar Gibbs free
energy, molar volume and isothermal compressibility in a first-order phase transition (D–F).
Schematic representation of the temperature dependence of the molar Gibbs free energy on the
temperature, molar entropy, and molar heat capacity in a second-order phase transition (G–I).

Paul Ehrenfest (1830–1933) proposed a classification of phase transitions according to which the **order of the phase transition** is defined by the lowest derivative
of the (Gibbs or Helmholtz) free energy that has a discontinuity upon crossing the
coexistence curve. Thus, **first-order** phase transitions are those for which one of the
first derivatives of the free energy is discontinuous (Figure 7.4 A–F), and **second-
order** phase transitions are those for which both of the first derivatives of the free
energy are continuous but at least one of the second derivatives is a discontinuous
function (Figure 7.4 G–I). In first-order phase transitions, c_P and k_T go to infinity
when the transition point is approached from either side (Figure 7.4 C, F), while in
second-order phase transitions the kink in the entropy results in a step in its derivative
(Figure 7.4 I). The infinite value of c_P results from the fact that a nonzero transfer
of energy as heat causes no change in the temperature as one phase is converted
to another. As we shall see in the next section, this amount of energy is the **latent
heat**.

Think about it...
Why does the isothermal compressibility k_T go to infinity at the liquid-gas coexistence region?

Answer

The infinite value of k_T results from the fact that a finite change in volume occurs at constant pressure P_0 when moving from the liquid to the gas phase along the coexistence region.

The order of a phase transition must be experimentally determined by performing accurate measurements of the compressibility and heat capacity, and check if these quantities become infinite at the phase transition. Second-order phase transitions are rather unusual. One well established example is the transition from normal to superconducting state.

The coexistence region in the (T, P) projection (dashed line in Figure 7.2 B) is enclosed by a line that is concave downwards, with a critical point at the top. In his classical theory of phase transitions Lev Landau (1908–1968) argued that the shape of this line, and, in particular, the existence of a critical point, results from the dependence of the molar free energy on the molar volume along the liquid-vapour coexistence line. In particular, according to Landau's theory, for a given $T = T(P)$ in the coexistence line, the free energy expressed as a function of v, $G(v)$, has two minima separated by a barrier, one corresponding to the liquid phase and another to the gas phase. The barrier becomes progressively smaller as the critical point is approached, with the two minima actually merging into a broad minimum at the critical point (Figure 7.5). In light of Landau's theory, a phase transition is called

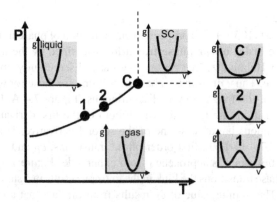

Figure 7.5   The shape of the free energy $g(v)$ along the liquid-gas coexistence line. The free energy barrier that separates the liquid and the gas phase becomes progressively smaller as the critical point is approached, actually vanishing at the critical point. Adapted from Dill and Bromberg (2002).

first-order when the free energy has two minima separated by a barrier, and higher-order when it has no barrier.

It is also common to classify phase transitions as first-order or **continuous**. Continuous phase transitions are those for which the second, or higher-order derivative of the free energy shows a discontinuity.

7.6   CLAUSIUS-CLAPEYRON EQUATION AND LATENT HEAT

The discontinuous change in entropy that occurs in a first-order phase transition is associated with the existence of a quantity termed **latent heat**. The latter is the amount of energy transferred as heat that is necessary to make a thermodynamic system change state without changing its temperature.

In what follows we will derive an equation that relates the slope of the liquid-gas coexistence line with latent heat for a one-component thermodynamic system. A similar equation exists for the other coexistence lines and the derivation is identical. The equation that describes the slope of the coexistence lines is called the Clausius-Clapeyron equation.

As we already know, a system composed of two phases (such as the gas phase g, and the liquid phase l) is in equilibrium if $T_g = T_l$, $P_g = P_l$, and $\mu_g = \mu_l$. Moreover, we also know from the Gibbs-Duhem equation (6.10) that T, P, and μ are not all independent. In particular, we can write $\mu = \mu(T,P)$. Thus, the condition of thermodynamic equilibrium along the liquid-gas coexistence line reduces to

$$\mu_l(T,P) = \mu_v(T,P).\tag{7.2}$$

Taking the differential of the equation above one gets

$$\left(\frac{\partial \mu_l}{\partial T}\right)_P dT + \left(\frac{\partial \mu_l}{\partial P}\right)_T dP = \left(\frac{\partial \mu_v}{\partial T}\right)_P dT + \left(\frac{\partial \mu_v}{\partial P}\right)_T dP.\tag{7.3}$$

For a one-component system $dg = d\mu$ with $dg = -sdT + vdP$. Thus

$$\left(\frac{\partial \mu}{\partial T}\right)_P = -s,$$

and

$$\left(\frac{\partial \mu}{\partial P}\right)_T = v.$$

Using the two last equations, we can rewrite equation (7.3) as

$$-s_l dT + v_l dP = -s_g dT + v_g dP,$$

and rearranging one obtains

$$\frac{dP}{dT} = \frac{s_g - s_l}{v_g - v_l}.\tag{7.4}$$

Recall from equation (6.4) that for a one-component system $\mu = g$. Since $g = h - Ts$ it is true that

$$\mu = h - Ts.$$

On the other hand, along the coexistence line

$$\mu_g = \mu_l.$$

Thus, along the coexistence line the following equality holds:

$$h_l - T s_l = h_g - T s_g,$$

where T is the **coexistence temperature**.

Rearranging the last equation one gets

$$s_g - s_l = \frac{h_g - h_l}{T}. \tag{7.5}$$

Finally, substituting (7.5) in (7.4) we obtain the **Clausius-Clapeyron** equation:

$$\frac{dP}{dT} = \frac{h_g - h_l}{T(v_g - v_l)} \tag{7.6}$$

The quantity $l = (h_g - h_l)$ in equation (7.6) is **the latent heat of vaporisation per particle from liquid to gas**. Alternatively, taking into account equation (7.5), the latent heat can be defined as

$$l = T(s_g - s_l). \tag{7.7}$$

The entropy and volume of the gas phase are typically larger than those of the liquid phase; the same is true for the molar volume and molar entropy. Consequently, the slope of the liquid-gas coexistence line is positive.

Think about it...

Consider a piece of ice that sits on the top of a table. What can you say about the ice's temperature?

Answer

Even if the ice cube (or part of it) had melted, its temperature could yet be 0°C due to latent heat. Since the ice has not melted one can say for sure that its temperature is lower than the table's temperature.

To obtain the Clausius-Clapeyron equation for the solid-liquid, and for the solid-gas coexistence lines one uses the same procedure as above but starting from the corresponding equilibrium condition. Namely,

$$\mu_s(T,P) = \mu_l(T,P), \tag{7.8}$$

and

$$\mu_s(T,P) = \mu_g(T,P). \tag{7.9}$$

In the particular case of water, the slope of the solid-liquid coexistence curve is, contrary to that of most substances, negative. Thus, for water, a higher pressure corresponds to a lower melting temperature. The reason for this *strange* behaviour of water is related to the network of hydrogen bonds between the water molecules, which causes them to be closer to each other in the liquid phase than in the ice phase. As a result, the molar volume of the liquid phase is smaller than that of the solid phase, and ice floats on liquid water.

Think about it...

Can you use the Clausius-Clapeyron for the melting line of water to try to explain the movement of glaciers?

Answer

Contrary to most substance, water exhibits a decrease of melting temperature with increasing pressure. In principle, one could argue that the bottom of a glacier experiences a very high pressure as a result of its weight, which can lead to some ice melting and forming a liquid layer of water that facilitates the movement of the glacier.

7.7   THE VAN DER WAALS GAS MODEL

The isotherm of the ideal gas model does not exhibit a phase transition. Microscopically, this stems from the lack of intermolecular interactions between the ideal gas particles. While nowadays it is well-known that non-charged particles interact with each other through pairwise forces derived from an intermolecular potential as the one represented in equation (1.3), things were not this clear back in the 19th century.

The Dutch physicist Johannes van der Waals (1837–1923) was the first to consider the effect of intermolecular interactions in the behaviour of a gas, although he was not absolutely sure about the nature and origin of those interactions. Basically, he imagined that each particle attracts all the others, within a fixed range of influence. He also reasoned that since liquids strongly resist compression, it should be because the constituent particles, which are also the ones present in the gas phase, must keep a minimum distance between them. The van der Waals gas is the first model of a real gas that takes into account the size of the gas particles, and the existence of forces between them. It is the basis of the van der Waals pressure equation of state, which granted van der Waals the Nobel prize in Physics in 1910. In what follows we will make a phenomenological derivation of the van der Waals equation of state. The reader should bear in mind that the van der Waals model provides a qualitative description of a simple fluid. This means that while it is able to replicate several physical features of its behaviour, it is not able to replicate the corresponding experimental values.

Consider a monoatomic gas formed by N identical particles that interact in a pairwise manner. The gas particles are represented as rigid spheres, i.e., impenetrable spheres that cannot overlap in space. Each sphere has radius r. Let P and V be the *measured* (or *observed*) pressure and volume of a real gas, respectively. We take as a starting point the ideal gas equation of state

$$P'V' = Nk_BT, \qquad (7.10)$$

where P' and V' are the pressure and volume of the ideal gas. In what follows we start by considering the effect of the size of the particles to correct V'; in doing so we will estimate V. Subsequently, we will consider the effect of the existence of intermolecular forces to correct P' and estimate P.

Since the rigid spheres cannot interpenetrate each other, the distance between two particles cannot be smaller than $\sigma = 2r$ (Figure 7.6 A). Therefore, to each pair of interacting particles one can associate an **excluded volume**

$$V_{pair} = \frac{4}{3}\pi(2r)^3 = \frac{4}{3}\pi\sigma^3.$$

Thus, the excluded volume per particle is

$$V_{partcle} = \frac{V_{pair}}{2}.$$

Considering that the gas is formed by N particles, the total excluded volume is

$$V_{excluded} = Nb,$$

with

$$b = \frac{2}{3}\pi\sigma^3$$

being a parameter characteristic of each specific particle. Based on these simple considerations one can estimate the measurable volume as being

$$V = V' + Nb. \qquad (7.11)$$

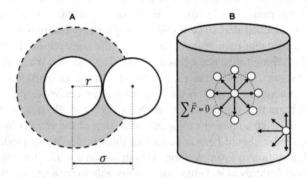

Figure 7.6 Excluded volume (in light-grey) for a pair of rigid particles. The minimum distance between the two particles is $\sigma = 2r$ (A). Pull force on particles located on the surface of the container (B).

Let us now turn our attention to the estimation of the pressure P of a real gas. In order to do so, we assume that the particles do not interact with the surface of the container. Since the forces between particles are isotropic, the resultant force acting on each particle in the bulk is null. On the other hand, a particle located on the surface of the container will be acted upon by a total non-null attractive force that will pull the particle towards the bulk (Figure 7.6 B). This *pull* towards the bulk, acting on the particles colliding with the surface per unit area and per unit time, results into P being smaller by an amount ΔP than the pressure P' corresponding to an ideal gas:

$$P = P' - \Delta P.$$

The number of particles that collide with the surface is proportional to the gas density, ρ. Moreover, according to van der Waals, the net attractive force acting on any particle in any direction is proportional to the density of molecules within its range of influence. Therefore, the net attractive force acting on the particles located on the surface will be proportional to ρ^2. Thus, the correction of the pressure due to the pull must be proportional to ρ^2:

$$\Delta P = a \left(\frac{N}{V} \right)^2,$$

where the parameter a is also specific of each particle type.

The pressure of a real gas can then estimated to be,

$$P = P' - a \left(\frac{N}{V} \right)^2. \tag{7.12}$$

Using equations equations (7.11) and (7.12) to respectively obtain P' and V', and inserting the corresponding expressions in (7.10) one finally obtains the **van der Waals pressure equation of state**:

$$P = \frac{N k_B T}{(V - Nb)} - a \left(\frac{N}{V} \right)^2 \tag{7.13}$$

Equation (7.13) is sometimes presented in the following form

$$\frac{P}{k_B T} = \frac{\rho}{1 - b\rho} - \frac{a\rho^2}{k_B T} \tag{7.14}$$

with $\rho = N/V$.

The isotherms corresponding to the van der Waals equation are represented in Figure 7.7 A. We note, in particular, that below the critical temperature, the isotherm shows a region of non-physical behaviour. Indeed, one observes that an increase in the volume results into a pressure increase leading to a negative value of the isothermal compressibility, k_T (5.33). This happens in the part of the curve that corresponds to the coexistence region observed experimentally for a real fluid. However, at point

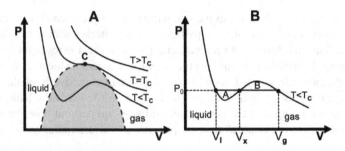

Figure 7.7 The isotherms predicted by the van der Waals equation of state (A), and the Maxwell construction (B). Below the critical temperature the van der Waals isotherm exhibits a zone of non-physical behaviour where the pressure increases with increasing volume.

(P_0, v_l) the derivative of the pressure with the volume is rather large, which is in line with what is expected for a liquid. On the other hand, at point (P_0, v_g), the derivative is rather small, which is typical of a gas. Furthermore, above the critical temperature T_C the van der Waals equation correctly describes the behaviour of a real fluid. Thus, the van der Waals isotherm does not correctly capture the coexistence region, but it is able to predict the existence of the gas phase, the liquid phase, as well as the critical and supercritical isotherms. Can one use the van der Waals equation to establish the pressure at which the two phases coexist? The answer to this question is *yes* and can be obtained by the so-called Maxwell construction, which we discuss in the next section.

7.7.1 MAXWELL CONSTRUCTION

Let us consider a van der Waals isotherm below T_C. Since T is constant along the isotherm, by using the Gibbs-Duhem equation (6.10) one can write

$$\left(\frac{\partial \mu}{\partial P}\right)_T = v,$$

which is equivalent to

$$d\mu = vdP, \text{ with constant } T.$$

Since

$$vdP = d(Pv) - Pdv,$$

it follows that

$$d\mu = d(Pv) - Pdv.$$

By integrating the equation above between v_l and v_g one obtains

$$\mu_g - \mu_l = \int_{v_l}^{v_g} [d(Pv) - Pdv]$$

$$= P_g v_g - P_l v_l - \int_{v_l}^{v_g} Pdv. \tag{7.15}$$

Since the system is in thermodynamic equilibrium in the coexistence region, the condition $\mu_l = \mu_g$ must hold. Moreover, along coexistence region $P_l = P_g = P_0$. Thus equation (7.15) simplifies to:

$$P_0(v_g - v_l) = \int_{v_l}^{v_g} P dv.$$

Considering a point v_x such that $v_l < v_x < v_g$ (Figure 7.7 B), the last equation can be rewritten as

$$P_0(v_x - v_l) + P_0(v_g - v_x) = \int_{v_l}^{v_x} P dv + \int_{v_x}^{v_g} P dv, \qquad (7.16)$$

and rearranging (7.16) one obtains

$$P_0(v_x - v_l) - \int_{v_l}^{v_x} P dv = \int_{v_x}^{v_g} P dv - P_0(v_g - v_x). \qquad (7.17)$$

The left hand side of equation (7.17) corresponds to area **A** in Figure 7.7 B, while the right hand side corresponds to area **B**.

Thus, in order for the equality in (7.17) to hold, it is necessary to ensure that the graphical condition

area **A** = area **B**

is satisfied, which will only happen for a certain value of the vapour pressure P_0. In other words, P_0 is the pressure for which the two areas are equal. This geometric analysis of the van der Waals isotherm that predicts the value of P_0 is known as the **Maxwell construction**. The latter allows one to conclude that while the van der Waals model does not correctly predict the occurrence of the liquid-gas coexistence region, it is nevertheless able to provide the pressure of phase coexistence.

The fact that the van der Waals model is able to qualitatively predict the super-critical isotherm ($T > T_C$), the critical isotherm ($T = T_C$), the existence of liquid and gas phases, together with the vapour pressure ($T < T_C$) is quite remarkable given its simplicity. This makes the van der Waals model an important step towards the understanding of phase transitions. In what follows we analyse in detail the critical isotherm.

7.7.2   CRITICAL ISOTHERM AND CORRESPONDING STATES

As previously mentioned, the liquid-gas coexistence line terminates at the so-called **critical point**. At the critical point the volume of the gas phase coincides with that of the liquid phase, i.e., there is no distinction between the two phases and the latent heat vanishes. Liquid and gas become a supercritical fluid. The volume changes continuously in moving from the liquid to the gas phase across the critical point. It is experimentally observed that at the critical point the isothermal compressibility and the heat capacity are both infinite. However, there is no sudden jump to an infinite value at one point as it happens below the critical temperature. They both rise smoothly towards an infinite value at the critical point. As we will see, this kind of phase transition is generally associated with *anomalous* phenomena.

The van der Waals equation of state predicts the existence of a critical isotherm as the one depicted in Figure 7.7 A. The critical isotherm does not possess an horizontal part as the isotherms below T_C, which is consistent with the fact that there is no distinction between the liquid and the gas phase. However, the slope of the isotherm is still zero at the critical point, indicating that the pressure does not change with volume. On the other hand, the isotherm's curvature changes sign at the critical point. From a mathematical standpoint the critical point is a **stationary inflection point** of the critical isotherm. This means that not only the first derivative of the pressure with respect to volume is zero

$$\left(\frac{\partial P}{\partial v} \right)_{T_C} = 0, \tag{7.18}$$

but that the second derivative is zero as well

$$\left(\frac{\partial^2 P}{\partial v^2} \right)_{T_C} = 0. \tag{7.19}$$

At the critical point it is not only the first and the second derivatives of the pressure that vanish. Indeed, by looking into the (v, T) projection of the phase diagram of the van der Waals fluid it is possible to see that the same happens with the first and second derivatives of the temperature,

$$\left(\frac{\partial T}{\partial v} \right)_{P_C} = \left(\frac{\partial^2 T}{\partial v^2} \right)_{P_C} = 0. \tag{7.20}$$

Thermodynamics breaks down at the critical point because, as previously mentioned and now illustrated by the van der Waals equation of state, the heat capacities together with other measurable thermodynamic coefficients (such as the thermal pressure, isothermal compressibility etc.) that are generally used to quantitatively characterise a thermodynamic system, become either zero or infinite at the critical point.

By using equation (7.13) to compute (7.18) and (7.19) one obtains an expression for the critical volume

$$v_C = 3b, \tag{7.21}$$

and for the critical temperature

$$k_B T_C = \frac{8a}{27b} \tag{7.22}$$

expressed as a function of parameters a and b.

By substituting V_C and T_C in (7.13) one gets the critical pressure

$$P_C = \frac{a}{27b^2}. \tag{7.23}$$

The critical parameters P_C, V_C, and T_C allow one to define a parameter, Z_C, termed **universal compressibility ratio,** which measures how much a van der Waals gas deviates from the ideal gas ($Z_C \to 1$)

$$Z_C = \frac{P_C v_C}{k_B T_C} = \frac{3}{8} = 0.375. \tag{7.24}$$

In general, it is observed that real gases deviate more from the ideal gas limit than as predicted from the van der Waals theory (e.g. in the case of water $Z_C = 0.226$), indicating that they are not perfectly captured by the van der Waals model.

The critical parameters allow defining the so-called **reduced variables**:

$$\overline{P} \equiv \frac{P}{P_C}, \tag{7.25}$$

$$\overline{T} \equiv \frac{T}{T_C}, \tag{7.26}$$

$$\overline{V} \equiv \frac{V}{V_C}, \tag{7.27}$$

and

$$\overline{\rho} \equiv \frac{\rho}{\rho_C}. \tag{7.28}$$

In equation (7.28), $\rho_C = v_C/v$ is the critical density.

If $\overline{P} = \overline{T} = \overline{V} = 1$ the system is in the critical state. If $\overline{P} > 1$ and $\overline{T} > 1$ the system is above critical conditions. Alternatively, it will be in subcritical conditions when $\overline{P} < 1$ and $\overline{T} < 1$.

The van der Waals equation expressed in reduced variables becomes

$$\overline{P} = \frac{8\overline{T}}{\left(3\overline{V} - 1\right)} - \frac{3}{\overline{V}^2}. \tag{7.29}$$

The reduced form of the van der Waals equation of state is somehow *universal* in the sense that the coefficients a and b, which are specific of each substance, are no longer there. All substances obey the same equation of state in terms of reduced variables. Reduced variables determine the so-called corresponding states. The **principle of corresponding states**, as originally established by van der Waals in 1880, states that:

At a given value of the reduced volume, $\overline{V}$, and reduced temperature, $\overline{T}$, all gases have the same reduced pressure, $\overline{P}$

Van der Waals conjectured that the principle of corresponding states is independent of the equation of state. It is viewed as one of the most – if not the most – useful contribution of van der Waals theory.

7.8   GUGGENHEIM CURVE AND CRITICALITY

In 1945, Edward Guggenheim (1901–1970) considered eight different substances (Ne, Ar, Kr, Xe, N_2, O_2, CO, CH_4) and studied experimentally the liquid-gas phase transition. The **Guggenheim curve** (Figure 7.8) shows that when plotted in reduced variables ($\overline{\rho}$ and $\overline{T}$), the coexistence curves of the eight different substances superpose remarkably well in the vicinity of the critical point, showing that the reduced

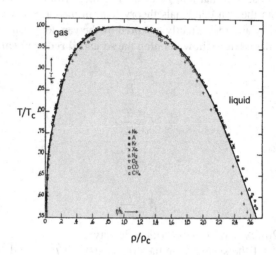

Figure 7.8 Guggenheim curve. Liquid-gas coexistence curve for eight different substances showing the reduced temperature as a function of reduced density. The solid line is the fit to a cubic function with a singularity of the form $(\rho - \rho_C) \propto (-t)^{\beta}$ with $t = (T - T_C)/T_C$ and $\beta = 0.33$. Adapted from Guggenheim (1945).

densites of the liquid phase and gas phase at the coexistence region depend on the reduced temperature in a *universal* manner. This is consistent with van der Waals' principle of corresponding states.

Actually, as the temperature approaches T_C, the difference between the densities of the liquid phase and gas phase, $(\rho_l - \rho_g)$, approaches zero indicating that it is no longer possible to distinguish between the two phases. At the critical point, the density of liquid phase and of the gas phase become the same, and the difference between the two phases disappears. As previously mentioned, they become a single, supercritical fluid. In the context of the study of phase transitions the quantity $(\rho_l - \rho_g)$ is termed **order parameter**. Order parameters are quantities which are non-zero below the critical temperature and zero above it.

7.9 CRITICAL EXPONENTS

The macroscopic properties of a thermodynamic system near the critical point are determined by the temperature. More precisely, they are a function of the amount by which the temperature deviates from the critical temperature. For this reason it is convenient to define the temperature in such a way that all critical points are equivalent. In general, the dimensionless parameter

$$t \equiv (T - T_C)/T_C \tag{7.30}$$

is used to study critical behaviour.

Experiments show that near the critical point thermodynamic properties are often proportional to $(-t)$ raised to some power, i.e., they are determined by **power-laws**

$$X \sim (-t)^k,$$

with X denoting some thermodynamic property, and k being the **critical exponent**. For example, in the case of the Guggenheim curve it is observed that

$$(\rho - \rho_C) \propto (-t)^\beta. \tag{7.31}$$

with $\beta = 0.33$ being the critical exponent. In what follows we briefly discuss the concept of critical exponent and why these quantities are important.

Critical exponents are the entities used to characterise critical phenomena, and the **theory of critical phenomena** is the theory behind the calculation or prediction of critical exponents. Some examples of power laws describing critical behaviour are

$$(\rho_l - \rho_g) \sim (-t)^\beta \quad (T < T_C) \qquad\qquad 0.3 < \beta < 0.5 \tag{7.32}$$
$$(v_l - v_g) \sim (P - P_C)^\delta \quad (P > P_C) \qquad\qquad 4.0 < \delta < 5.0 \tag{7.33}$$
$$C_V \sim t^{-\alpha} \quad (T > T_C) \qquad\qquad -0.2 < \alpha < 0.3 \tag{7.34}$$
$$C_V \sim (-t)^{-\alpha'} \quad (T < T_C) \qquad\qquad 0 < \alpha' < 0.2 \tag{7.35}$$
$$k_T \sim t^{-\gamma} \quad (T > T_C) \qquad\qquad 1.2 < \gamma < 1.4 \tag{7.36}$$
$$k_T \sim (-t)^{-\gamma'} \quad (T < T_C) \qquad\qquad 1.0 < \gamma' < 1.2 \tag{7.37}$$

with the **critical exponents** being represented by β, δ, α, α', γ, and γ'.

As shown in the equations above, the values of critical exponents are strongly conserved among different fluid systems (e.g. the eight substances represented in the Guggenheim curve all share the same critical exponent $\beta = 0.33$), and even so among physical systems that are clearly different (e.g. binary mixtures, metal alloys, and ferromagnets). This remarkable observation is another statement of *universality*.

A property of power laws is that they are **scale invariant**. Given a function $f(t) = \alpha t^k$, and scaling the argument by a constant factor c one obtains $f(ct) = \alpha(ct)^{-k} = \alpha c^{-k} f(t) \propto f(t)$. Thus, it follows that all power laws with a particular scaling exponent are equivalent up to constant factors, since each one is simply a scaled version of the others.

In the particular case of a fluid system at the critical point the density has fluctuations of *all* length scales, from microscopic to macroscopic ones, which are about the systems' size. Since there is not a preferred density scale, the system is scale invariant. In particular, there are density fluctuations whose size is comparable to the wavelength of light, and scattered light causes the normally transparent liquid to become *opalescent*, exhibiting a milky appearance. This so-called **critical opalescence** was firstly reported by Charles Cagniard de la Tour (1777–1859) in 1823 for mixtures of alcohol and water, but its importance was only recognised by Thomas Andrews (1813–1885) in 1869 after experiments on the liquid-gas transition of carbon dioxide. It is therefore one of the first investigated manifestations of critical phenomena.

The van der Waals equation of state is able to qualitatively predict critical behaviour but it is not able to correctly predict the experimental values of the critical exponents (e.g. it predicts $\beta = 0.5$ instead of 0.33). A successful treatment of critical phenomena is one that is able to quantitatively predict the critical exponents by properly treating fluctuations. It was developed by Ken Wilson (1936–2013), and is called **Wilson's renormalisation group method**. For his achievements, Wilson received the Nobel Prize in Physics in 1971. The interested reader is referred to the classic textbook by Eugene Stanley, and to the more recent textbook by Nishimori and Ortiz to learn more about criticality and the physical theory behind it.

7.10   LEARNING OUTCOMES

At the end of this chapter the reader should be able to:

1. Know the concept of phase and understand what is a phase diagram.
2. Be able to analyse the phase diagram of a simple fluid. In particular, to understand that the (T,P) projection provides the coexistence lines, and that the (V,P) projection provides the isotherms. To know what is the triple point, and that the liquid-gas coexistence line terminates at the critical point.
3. Know the Gibbs phase rule.
4. Know what is a phase transition, and how to classify phase transitions.
5. Derive the Clausius-Clapeyron equation; understand that it provides the slope of a coexistence line, which is related with latent heat.
6. Know that the van der Waals model is the first proposed model of a real gas and understand why it is important.
7. Be able to derive the van der Waals pressure equation of state by including excluded volume and intermolecular interactions.
8. Be able to critically analyse the van der Waals model by understanding its virtues and pointing out its limitations.
9. Use the Gibbs-Duhem equation to derive the Maxwell construction and understand that it provides the pressure vapour as predicted by the van der Waals model.
10. Know what are reduced variables, the principle of corresponding states, and understand its importance.
11. Understand that the Guggenheim curve is a manifestation of the principle of corresponding states.
12. Know that critical exponents describe the behaviour of a physical system in the critical point.

7.11   WORKED PROBLEMS

PROBLEM 7.1
Determine the internal energy of a van der Waals gas by considering an isothermal compression of an ideal gas.

Solution

In this process N is fixed and we take $U = U(T,V)$.
Then

$$dU = \left(\frac{\partial U}{\partial T}\right)_V dT + \left(\frac{\partial U}{\partial V}\right)_T dV.$$

By performing an isothermal gas compression ($dT = 0$) from an initial state (U_i, V_i) to a final state with (U_f, V_f) the previous equation can be rewritten as

$$U_f - U_i = \int_{V_i}^{V_f} \left(\frac{\partial U}{\partial V}\right)_T dV, \text{ constant } T.$$

According to Problem (5.5)

$$\left(\frac{\partial U}{\partial V}\right)_T = T\left(\frac{\partial P}{\partial T}\right)_V - P.$$

By differentiating the van der Waals pressure equation (7.13) it comes that

$$\left(\frac{\partial P}{\partial T}\right)_V = \frac{Nk_b}{V - Nb} = \frac{1}{T}\left(P + \frac{N^2 a}{V^2}\right),$$

so that

$$\left(\frac{\partial U}{\partial V}\right)_T = \frac{N^2 a}{V^2},$$

and therefore

$$U_f - U_i = N^2 a \int_{V_i}^{V_f} \frac{dV}{V^2},$$

In the limit $V_i \to \infty$, the real gas approaches an ideal gas. Choosing $V_i = \infty$ fixes the integration constant $U_i = U_{ideal}$. Taking $U_f = U_{vdW}$ and $V_f = V$ we can thus write

$$U_{vdW} = U_{ideal} - \frac{N^2 a}{V},$$

with $U_{ideal} = \frac{3}{2} N k_B T$.

PROBLEM 7.2
Based on the result of the previous problem determine the critical exponent
α.

Solution

The critical exponent α describes the behaviour of the heat capacity at constant volume in the vicinity of the critical point according to equations (7.35) and (7.36). Using the definition of $C_V = \left(\frac{\partial U}{\partial T}\right)_V$ it comes that for the van der Waals fluid

$$C_V = \frac{3}{2} N k_B = t^0 \frac{3}{2} N k_B.$$

Thus, for the van der Waals fluid $\alpha = \alpha' = 0$.

PROBLEM 7.3
Consider $S = S(T,V)$ and compute the entropy change of an expansion process of a van der Waals gas of N particles from state (T_1, V_1) to state (T_2, V_2).

Solution

Since $S = S(T,V)$,

$$dS = \left(\frac{\partial S}{\partial T}\right)_V dT + \left(\frac{\partial S}{\partial V}\right)_T dV.$$

Using the equality expressed by equation (3.30),

$$\left(\frac{\partial S}{\partial T}\right)_V = \frac{C_V}{T},$$

and the Maxwell relation expressed by equation (5.26),

$$\left(\frac{\partial S}{\partial V}\right)_T = \left(\frac{\partial P}{\partial T}\right)_V,$$

it comes that

$$dS = \frac{C_V}{T}dT + \left(\frac{\partial P}{\partial T}\right)_V dV$$
$$= \frac{C_V}{T}dT + \frac{R}{V-Nb}dV.$$

Integrating the equation above, first at constant temperature from state (T_1, V_1) to state (T_1, V_2):

$$S(T_1, V_2) = S(T_1, V_1) + \int_{V_1}^{V_2} \left(\frac{\partial S}{\partial V}\right)_T dV$$
$$= S(T_1, V_1) + \int_{V_1}^{V_2} \frac{R}{V-Nb}dV$$
$$= S(T_1, V_1) + R\ln\left(\frac{V_2 - Nb}{V_1 - Nb}\right),$$

and then at constant volume from state (T_1, V_2) to state (T_2, V_2):

$$S(T_2, V_2) = S(T_1, V_2) + \int_{T_1}^{T_2} \left(\frac{\partial S}{\partial T}\right)_V dT$$
$$= S(T_1, V_2) + \int_{T_1}^{T_2} \frac{C_V}{T}dT$$
$$= S(T_1, V_2) + C_V \ln\left(\frac{T_2}{T_1}\right)$$
$$= S(T_1, V_1) + R\ln\left(\frac{V_2 - Nb}{V_1 - Nb}\right) + C_V \ln\left(\frac{T_2}{T_1}\right),$$

where we have taken into account the fact that for a van der Waals gas, C_V does not depend on the temperature (see problem 7.2). Thus

$$S(T_2, V_2) - S(T_1, V_1) = R\ln\left(\frac{V_2 - Nb}{V_1 - Nb}\right) + C_V \ln\left(\frac{T_2}{T_1}\right).$$

7.12 SUGGESTED PROBLEMS

PROBLEM 7.4
Determine the number of degrees of freedom for a two-component liquid mixture in equilibrium with its gas.

PROBLEM 7.5
In a two-component system, what is the maximum number of phases that can exist in equilibrium?

PROBLEM 7.6
At atmospheric pressure, silver melts at $T = 1235$ K and its volume expands about 4%, the actual volume change along melting being about 0.4 cm^3/mol. Its latent heat of fusion is 11.950 J/mol. How much must the pressure increase to raise its melting point by 1 K?

PROBLEM 7.7
Derive an approximate expression for the Clausius-Clapeyron equation by taking into account that the latent heat Δh is a positive constant, the molar volume of the gas is much larger than that of the liquid, $v_g - v_l \approx v_g$, and that $v_g \approx \frac{RT}{P}$.

PROBLEM 7.8
Consider the van der Waals equation as given by (7.13). Evaluate P_C, T_C, and V_C.

PROBLEM 7.9
Consider the van der Waals equation expressed in molar volume. Show that it can be written as a cubic equation in v

$$Pv^3 - (Pb + RT)v^2 + av - ab = 0,$$

and that at the critical point (P_C, v_C, T_C) the equation above reduces to

$$(v - v_C)^3 = 0.$$

PROBLEM 7.10
It is observed that near the critical point the dependence of pressure on density is

$$\frac{P}{k_B T} = \frac{P_c}{k_B T_c} + C(\rho - \rho_C)^\delta,$$

where C is a constant. What does the van der Waals theory predict for δ?

Hint: Expand the equation of state about the critical density and temperature.

PROBLEM 7.11
**Consider the definition of isothermal compressibility, k_T. What does the van
der Waals theory predict for γ?**

REFERENCES

1. Callen, H. B. (1960). Thermodynamics. Wiley.

2. Dill, K. A. & Bromberg S. (2002). Molecular Driving Forces: Statistical Thermodynamics in Biology and Chemistry. Taylor & Francis.

3. Guggenheim, E. A. (1945) The principle of corresponding states. J. Chem. Phys. 13: 253–261.

4. Kondepudi, D. (2008). Introduction to Modern Thermodynamics: From Heat Engines to Dissipative Structures. Wiley.

5. Nishimori, H. and Ortiz G. (2011) Elements of Phase Transitions and Critical Phenomena. Oxford University Press.

6. Sekerka, R. F. (2015) Thermal Physics: Thermodynamics and Statistical Mechanics for Physicists and Engineers. Elsevier.

7. Stanley, H. E. (1971) Introduction to Phase Transitions and Critical Phenomena. Oxford University Press.

8. Wilson, K. G. (1979) Problems in physics with many scales of lenght. Scientific American 241:158-179.

8 Magnetic Systems

So far, we have been illustrating the theory of thermodynamics in the context of simple fluid systems. The present chapter is focused on a completely different system, that of a simple magnetic system. It starts by defining magnetic work done on the system and to establish the differential of the internal energy. It moves on by presenting the thermodynamic potentials for magnetic systems, and the equations of state for paramagnetic and diamagnetic systems. It ends up by discussing a particular cooling process termed adiabatic demagnetisation, and the occurrence of absolute negative temperature in systems that exhibit a phenomenon called population inversion.

8.1 INTRODUCTION

A **magnetic system** differs from a non-magnetic system because its internal energy depends on an external applied magnetic field; this is generally described as the system interacting with the magnetic field. The extensive property of the thermodynamic system associated with this interaction is the **magnetic moment**.

The magnetic moment first appears in classical electromagnetism, to describe the interaction between the magnetic field $\vec{H}$ and an electric current I circulating in a flat loop. If the area enclosed by the loop is A, the magnetic moment $\vec{m}$ of the current loop is defined as a vector with magnitude IA, perpendicular to the plane of the loop, pointing towards the side from which the current is seen to circulate counterclockwise. The (magnetic) energy associated with this interaction is

$$E = -\mu_0 \, \vec{m} \cdot \vec{H},$$

with μ_0 being the permeability of empty space. This concept generalises to any system interacting with a magnetic field, but, in most materials, the magnetic moment cannot be explained by an electric current flowing in its volume. It involves the contribution of atomic magnetic moments associated with the existence of electron and nucleon spins. These quantities can only be described in the scope of quantum theory, and the corresponding magnetic moments are considered an intrinsic property of matter.

In general, all materials respond to an applied magnetic field, even if in zero magnetic field their magnetic moment is zero. That is the case of **paramagnetic** and **diamagnetic** materials that only develop a magnetic moment in a non-zero magnetic field. Other materials may display a magnetic moment in the absence of external magnetic field. These are the ones generally designated **magnetic materials**, which include **ferromagnets**, **ferrimagnets**, and **antiferromagnets**, and correspond to materials having ordered arrangements of the atomic/electron magnetic moments (Figure 8.1).

Note that a difference is made in this text between magnetic system, which is a system whose internal energy depends on magnetic field, and magnetic material,

DOI: 10.1201/9781003091929-8

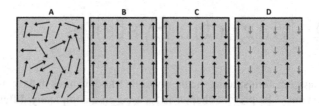

Figure 8.1   Organisation of the atomic magnetic moments inside a material, showing the most frequent three types of magnetic ordering: Paramagnetic (A), ferromagnetic (B), anti-ferromagnetic (C), and ferrimagnetic (D). Diamagnetic materials have zero atomic magnetic moments in zero field.

which is a material that exhibits magnetic moment in the absence of a magnetic field. The former is more general and includes the magnetic materials.

The magnetic state of a material is characterised by the **magnetisation** $\vec{M}$, de-fined as magnetic moment per unit volume (the molar magnetisation, magnetic moment per mol, or the mass magnetisation, magnetic moment per mass unit, can also be used).

Consider a region in empty space where a constant magnetic field $\vec{H}_{ext}$ is applied (the subscript ext indicates that it has no dependence on the system and must be taken as external). In that region we can characterise the magnetic field by $\vec{H}_{ext}$ or by the magnetic induction field $\vec{B}_0 = \mu_0 \vec{H}_{ext}$. If a material is placed in that region, the local relation between $\vec{B}$ and $\vec{H}$ inside the material is

$$\vec{B} = \mu_0 \left(\vec{H} + \vec{M} \right) = \mu \vec{H}, \tag{8.1}$$

which defines the **magnetic permeability** μ of the system (not to be confused with the chemical potential). Due to the homogeneous assumption, we will consider that the relation remains valid for any finite region of the thermodynamic system.

Each microscopic magnetic moment in the material creates a magnetic dipolar field. The addition of these fields for all the magnetic moments in the material re-sults in a magnetic field opposite to the external field, named **demagnetising field**. In general, this demagnetising field is not important for diamagnetic and paramag-netic systems since the magnetisation values are small compared with the magnetic field, but for magnetic materials both applied field and demagnetising field can be of the same order of magnitude, implying, in equation (8.1), a magnetic field value cor-rected for the demagnetising field. The demagnetizing field is, locally, proportional to the magnetisation, but it is in general non uniform.

In conclusion, placing a magnetic material in a magnetic field usually im-plies two important modifications inside the material: 1) the magnetic induction field changes, and 2) it can be non-uniform, varying from one region to other. The non-uniformity may be due to the demagnetising field and its dependence of the shape of the material or, for ordered states, it can be a consequence of the existence of different magnetic domains. (A **magnetic domain** corresponds to a region where the microscopic arrangement favours a specific alignment of the magnetic moments. In a magnetic material, different directions of alignment are possible due to the bonding states symmetry, allowing for the coexistence of several magnetic domains). Since we are studying macroscopic homogeneous systems, it will be assumed that the magnetic induction is constant within the system, which implies a material with a particular shape (the exact condition for a spatially uniform magnetic field is fulfilled for an ellipsoid in a uniform external field). We will also assume that, in the case where magnetic domains can form, that the system is contained in one single domain.

Think about it...

The validity of equation (8.1) for any region of a homogeneous system, requires that the magnetic permeability is constant?

Answer

Yes, since the system is homogeneous, any part of the system must verify the same relation for the intensive variable.

8.2   MAGNETIC WORK AND INTERNAL ENERGY

To deduce the internal energy term associated with the magnetic interaction, we will calculate the magnetic work supplied to the system when its magnetic moment changes under a magnetic field.

As source of the external magnetic field, we choose a long solenoid, whose length L is much larger than our system, with diameter $D \ll L$, through which flows an electric current I supplied by an electrical source (Figure 8.2 A).

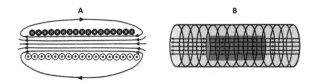

Figure 8.2   Solenoid in empty space showing the uniform field in the interior region (far from the solenoid ends) (A). System inside the solenoid occupying N' loops (B).

We will consider that the coil has N turns (each spanning an area A) and is fabricated with a **superconducting material**, so no power is dissipated when the constant electrical current I flows in the solenoid. Far from the edges, the geometry of the solenoid is such that (1) the magnetic field is approximately uniform inside the solenoid (in empty space $B = \mu_0 H = \mu_0 n I$ with $n = N/L$), and null outside and (2) the energy stored in the magnetic field is essentially the energy in the uniform region of the magnetic field.

The **electric work** supplied by the source to implement a magnetic field in empty space is

$$W = \int_0^t VI \, dt.$$

With no resistance, the potential difference V supplied by the source is only that required to overcome the electromagnetic field induced by the variation of magnetic flux $\varphi = BNA$:

$$W = \int_0^t \frac{\partial \varphi}{\partial t} I \, dt = N \int_0^t A \frac{\partial B}{\partial t} I \, dt \qquad (8.2)$$

Performing the change in integration variable $t \to H = B/\mu_0 = nI$,

$$W = NA \int_0^{H_{ext}} \mu_0 \frac{H}{n} \, dH = \mu_0 \, LA \int_0^{H_{ext}} H \, dH = \mu_0 \, \mathcal{V} \frac{H^2}{2}, \qquad (8.3)$$

where $\mathcal{V}$ denotes the volume enclosed by the solenoid.

With the system placed inside the solenoid (Figure 8.2 B), the work will be different. We will assume that the field in the system region is uniform and that the system does not influence the magnetic field outside its volume. The magnetic field is changed but only inside the system. Indeed, outside the system one has, as before:

$$\vec{B}_0 = \mu_0 \vec{H}_{ext} = \mu_0 n I \vec{u}, \qquad (8.4)$$

with $\vec{u}$ a unit vector along the axis of the solenoid, while inside the system one has:

$$\vec{B} = \mu_0 \left(\vec{H}_{ext} + \vec{M} \right) = \mu_0 \left(n I \vec{u} + \vec{M} \right). \qquad (8.5)$$

The electric work supplied by the source inside the solenoid,

$$W = \int_0^t VI \, dt = \int_0^t \frac{\partial \varphi}{\partial t} I \, dt,$$

can be rewritten as

$$W = \sum_{n'} \int_0^t A' \frac{\partial B_\parallel}{\partial t} I \, dt + \sum_n \int_0^t A'' \frac{\partial B_\parallel}{\partial t} I \, dt,$$

where $B_\parallel$ denotes the component of the magnetic induction field along the axis of the solenoid, the first sum covers the turns occupied by our system, filling an area A', and the second sum covers the turns over empty space. For simplicity we will assume that the system occupies the whole transverse section of the solenoid, and that it is much longer than wide (in this case the cylindrical shape approaches an ellipsoidal shape to assure homogeneity). Computing the sums and substituting $I = H/n$,

$$W = N'A \int_0^t \frac{\partial B_\parallel}{\partial t} \frac{H}{n} \, dt + (N - N')A \int_0^t \frac{\partial B_{0\parallel}}{\partial t} \frac{H}{n} \, dt$$

and changing variable in the first integral from t to B and in the second integral from t to H as before,

$$W = N' A \int_0^B \frac{H}{n} \, dB_\parallel + (N - N')A\mu_0 \int_0^H \frac{H}{n} \, dH.$$

Using (8.5), $dB_\| = \mu_0(dH + dM_\|)$, and we obtain

$$W = \mu_0 LA \int_0^H H\, dH + \mu_0 LA \frac{N'}{N} \int_0^M H dM_\| = V\mu_0 \int_0^H H\, dH + \mu_0 \int_0^M L'A\, H\, dM_\|.$$

Finally, since $L'A = V'$ is the volume occupied by the system inside the solenoid,

$$W = V\mu_0 \frac{H^2}{2} + \mu_0 \int_0^{\mathfrak{M}_\|} H\, d\mathfrak{M}_\|, \tag{8.6}$$

with

$$\mathfrak{M}_\| = V'M_\|$$

being the component of the magnetic moment of the system $\mathfrak{M}$ parallel to $\vec{H}$.

The first term in equation (8.6) is recognised as the work required to establish the magnetic field in the empty solenoid and does not depend on the system. So, we can identify the second term as **work done on the system** by the magnetic field $\vec{H}$:

$$W = \mu_0 \int_0^{\mathfrak{M}_\|} H d\mathfrak{M}_\|, \tag{8.7}$$

so that,

$$dW = \mu_0 H d\mathfrak{M}_\|. \tag{8.8}$$

Thus, we may conclude that the **internal energy of the magnetic system**, U, is a function of entropy, volume, number of particles, and magnetic moment component along the external magnetic field:

$$U = U\left(S, V, N, \mathfrak{M}_\|\right). \tag{8.9}$$

The differential dU is given by

$$dU = TdS - PdV + \mu\, dN + \mu_0 H d\mathfrak{M}_\| = TdS - PdV + \mu\, dN + \mu_0 \vec{H} \cdot d\vec{\mathfrak{M}}, \tag{8.10}$$

with

$$\left(\frac{\partial U}{\partial \mathfrak{M}_\|}\right)_{S,V,N} = \mu_0 H.$$

For magnetic systems it is common to write the internal energy in terms of the magnetisation:

$$\vec{M} = \frac{\vec{\mathfrak{M}}}{V},$$

or molar magnetisation,

$$\vec{M}_{mol} = \frac{N_{Av} \vec{\mathfrak{M}}}{N}.$$

In the former case, for constant volume processes:

$$du = Tds + \mu\, dn + \mu_0 \vec{H} \cdot d\vec{M}, \tag{8.11}$$

where $u = U/V$, $s = S/V$, and $n = N/V$.

Although we took the source of the magnetic field to be independent of the system to obtain this result, the same expression can be extended for systems other than diamagnetic and paramagnetic, if linearity is assumed between $\vec{B}$ and $\vec{H}$.

Using the energy density deduced from Maxwell equations:

$$u = \frac{1}{2}\left(\vec{B}\cdot\vec{H}\right) = \frac{\mu_0}{2}\left(\vec{M}\cdot\vec{H} + \vec{H}^2\right),\qquad(8.12)$$

the linearity between $\vec{B}$ and $\vec{H}$, written as $\vec{B} = \mu\vec{H}$, also implies a linear relation between the magnetisation and the magnetic field, $\vec{M} = \chi\vec{H}$, with χ being the **magnetic susceptibility** and $\mu = \mu_0(\chi + 1)$. Based on these considerations equation (8.12) can then be written as

$$u = \frac{\mu_0}{2}(\chi + 1)\vec{H}^2.$$

Antiferromagnetic systems and ferromagnetic systems at low fields obey the linearity condition, but $\vec{H}$ can be significantly different than $\vec{H}_{ext}$ due to the demagnetising field. This implies that subtracting the energy density associated with the external magnetic field production in empty space will not totally cancel the $\vec{H}^2$ term:

$$u - u_{empty\,space} = \frac{\mu_0}{2}\chi\vec{H}^2 + \frac{\mu_0}{2}\left(\vec{H}^2 - \vec{H}_{ext}^2\right)$$

and

$$du = \mu_0\vec{H}\cdot d\vec{M} + \mu_0\left(\vec{H}\cdot d\vec{H} - \vec{H}_{ext}\cdot d\vec{H}_{ext}\right).$$

This poses some formal questions concerning the convexity of the internal energy, defined in (8.11), but it is not considered relevant in general.

8.3 THERMODYNAMIC POTENTIALS FOR MAGNETIC SYSTEMS

As in the case of a simple fluid system, it is possible to define thermodynamic potentials for a simple magnetic system, sometimes referred to as **thermodynamic pseudo-potentials** due to the convexity issue question mentioned above.

Let us consider a one-component system with a fixed number of particles N and fixed volume V. In this case equation (8.10) simplifies to:

$$dU = TdS + \mu_0\vec{H}\cdot d\mathfrak{M}\qquad(8.13)$$

For transformations where the intensive parameters are the controlled variables, it will be useful to define the associated thermodynamic potentials using the Legendre transforms discussed in section 5.2 of Chapter 5.

The **magnetic enthalpy** $E_m = E_m(S,V,N,\vec{H})$ is defined as the Legendre transform of U with respect to $\mathfrak{M}$:

$$E_m = \mathcal{L}_{\mathfrak{M}}(U) = U - \mu_0\vec{H}\cdot\mathfrak{M},$$

and by using equation (8.13) it is easy to see that the differential of E_m is given by

$$dE_m = TdS - \mu_0 \vec{\mathfrak{M}} \cdot d\vec{H}. \tag{8.14}$$

The **magnetic free energy** $F_m = F_m(T,V,N,\vec{\mathfrak{M}})$ is defined as the Legendre transform of U with respect to S:

$$F_m = \mathcal{L}_S(U) = U - TS,$$

and the differential of F_m is

$$dF_m = -SdT + \mu_0 \vec{H} \cdot d\vec{\mathfrak{M}}. \tag{8.15}$$

Finally, the **magnetic free enthalpy** $G_m = G_m(T,V,N,\vec{H})$ can be defined as the Legendre transform of U with respect to both S and $\vec{\mathfrak{M}}$:

$$G_m = \mathcal{L}_{S,\vec{\mathfrak{M}}}(U) = U - TS - \mu_0 \vec{H} \cdot \vec{\mathfrak{M}},$$

and its differential is

$$dG_m = -SdT - \mu_0 \vec{\mathfrak{M}} \cdot d\vec{H}. \tag{8.16}$$

Note that the magnetic energy term of equation (8.7) is present in dF_m, but this term is replaced by $-\mu_0 \vec{\mathfrak{M}} \cdot d\vec{H}$ in dE_m and in dG_m.

Table 8.1 provides a summary of the four thermodynamic potentials, U, E_m, F_m, and G_m, for a closed single-component system with fixed volume, including their first derivatives.

Function	Differential	Natural variables	First derivatives	
U	$dU = TdS + \mu_0 \vec{H} \cdot d\vec{\mathfrak{M}}$	$(S,\mathfrak{M}_\parallel)$	$T = \left(\dfrac{\partial U}{\partial S}\right)_{\mathfrak{M}_\parallel}$	$H = \dfrac{1}{\mu_0}\left(\dfrac{\partial U}{\partial \mathfrak{M}_\parallel}\right)_S$
$E_m = U - \mu_0 \vec{H} \cdot \vec{\mathfrak{M}}$	$dE_m = TdS - \mu_0 \vec{\mathfrak{M}} \cdot d\vec{H}$	(S,H)	$T = \left(\dfrac{\partial E_m}{\partial S}\right)_H$	$\mathfrak{M}_\parallel = -\dfrac{1}{\mu_0}\left(\dfrac{\partial E_m}{\partial H}\right)_S$
$F_m = U - TS$	$dF_m = -SdT + \mu_0 \vec{H} \cdot d\vec{\mathfrak{M}}$	$(T,\mathfrak{M}_\parallel)$	$S = -\left(\dfrac{\partial F_m}{\partial T}\right)_{\mathfrak{M}_\parallel}$	$H = \dfrac{1}{\mu_0}\left(\dfrac{\partial F_m}{\partial \mathfrak{M}_\parallel}\right)_T$
$G_m = H_m - TS$	$dG_m = -SdT - \mu_0 \vec{\mathfrak{M}} \cdot d\vec{H}$	(T,H)	$S = -\left(\dfrac{\partial G_m}{\partial T}\right)_{\mathfrak{M}_\parallel}$	$\mathfrak{M}_\parallel = -\dfrac{1}{\mu_0}\left(\dfrac{\partial G_m}{\partial H}\right)_T$

Table 8.1
Differentials, natural variables, and first derivatives of the thermodynamic potentials that are useful to study closed, single-component magnetic systems with fixed volume.

Since dU, dE_m, dF_m, and dG_m are exact differentials, equations (8.13–8.16) establish the following Maxwell relations for a closed magnetic system where V is fixed:

$$\left(\frac{\partial T}{\partial \mathfrak{M}_\parallel}\right)_S = \mu_0 \left(\frac{\partial H}{\partial S}\right)_{\mathfrak{M}_\parallel} \tag{8.17}$$

$$\left(\frac{\partial T}{\partial H}\right)_S = -\mu_0 \left(\frac{\partial \mathfrak{M}_\parallel}{\partial S}\right)_H \tag{8.18}$$

$$\left(\frac{\partial S}{\partial \mathfrak{M}_\parallel}\right)_T = -\mu_0 \left(\frac{\partial H}{\partial T}\right)_{\mathfrak{M}_\parallel} \tag{8.19}$$

$$\left(\frac{\partial \mathfrak{M}_\parallel}{\partial T}\right)_H = \frac{1}{\mu_0} \left(\frac{\partial S}{\partial H}\right)_T \tag{8.20}$$

Let H_m be the enthalpy of a magnetic system (different from E_m), such that

$$H_m = U + PV.$$

For a closed magnetic system the differential of H_m is given by

$$dH_m = T dS + V dP + \mu_0 \vec{H} \cdot d\vec{\mathfrak{M}}. \tag{8.21}$$

At constant pressure, equation (8.21) is formally equivalent to equation (8.13). Therefore, a set of Maxwell relations equivalent to (8.17)–(8.20) can be obtained by considering equation (8.21), and the differentials of the Legendre transforms of $H_m = H_m\left(S, P, \mathfrak{M}_\parallel\right)$ with respect to S, P, and with respect to both S and P. This is left as an exercise for the reader.

8.4   THERMODYNAMIC COEFFICIENTS

The experimental study of magnetic systems, like all other systems, is carried out by studying how some thermodynamic variables depend on the others, using adequate physical processes. The derivatives that relate the rates of change of the different variables define thermodynamic coefficients that parametrise the behaviour of the systems. In section 5.5 of Chapter 5 we introduced several thermodynamic coefficients such as the coefficient of isobaric expansivity β_P (or α) (5.29), and the coefficient of isothermal compressibility k_T (5.33), and we also discussed the importance of the heat capacities at constant volume, C_V, and constant pressure, C_P. For closed magnetic systems, kept at constant pressure, there are other important coefficients, defined in a similar way, namely:

The **heat capacity at constant magnetic** field:

$$C_H \equiv T \left(\frac{\partial S}{\partial T}\right)_H. \tag{8.22}$$

The **heat capacity at constant magnetisation**:

$$C_M \equiv T \left(\frac{\partial S}{\partial T}\right)_M. \tag{8.23}$$

The **coefficient of thermal magnetisation**, associated with the dependence of magnetisation on temperature:

$$\alpha_H \equiv \frac{1}{V}\left(\frac{\partial \mathfrak{M}_\parallel}{\partial T}\right)_H . \tag{8.24}$$

And the two coefficients relating magnetic variables, namely, the **isothermal magnetic susceptance**

$$\kappa_T \equiv \frac{1}{V}\left(\frac{\partial \mathfrak{M}_\parallel}{\partial H_{ext}}\right)_T , \tag{8.25}$$

and the **isothermal magnetic susceptibility**:

$$\chi_T \equiv \left(\frac{\partial M_\parallel}{\partial H}\right)_T . \tag{8.26}$$

In homogeneous systems, where demagnetising field can be neglected $\kappa_T = \chi_T$ The four magnetic coefficients are not independent (see worked problem 8.3) and

$$C_M - C_H = -V\frac{\mu_0 T}{\chi_T}\alpha_H^2 . \tag{8.27}$$

8.5 EQUATIONS OF STATE

As mentioned before, magnetic systems can have different behaviours, and for each one different equations of state are defined. We discuss equations of state for paramagnetic and diamagnetic systems. Since linearity was assumed between M and H, all equations of state share this property.

8.5.1 DIAMAGNETIC SYSTEMS

A diamagnetic system only exhibits magnetic moment in an applied magnetic field, the magnetisation being proportional to the magnetic field but pointing in the opposite direction. The negative magnetic susceptibility is usually small (typically of the order 10^{-6} in SI units) and depends on the atomic density.

In the case of insulators, these materials are in general composed of particles, atoms or molecules, with zero magnetic moment. When a magnetic field is applied, opposite magnetic moments are induced at the particles positions, an effect that can be thought of classically as a manifestation of **Lenz's law**. The equation for the magnetisation is proportional to the density of particles and to the magnetic field:

$$M = \chi H = -A\frac{N}{V}H, \tag{8.28}$$

where A is a constant characteristic of the material. Since (8.28) involves an intensive variable, it can be considered an equation of state.

According to equation (6.29), for liquids and solids, the volume depends weekly on temperature and pressure as

$$V = V_0 \left(1 + \alpha \ (T - T_0) - k_T P \right),$$

where α (or β_P), the coefficient of isobaric expansivity (5.29), has values of the order of 10^{-4}, and k_T, the isothermal compressibility (5.33), has values of the order of 10^{-9} Pa^{-1}. T_0 and V_0 are the volume and temperature of a reference state. Including this dependence in the equation of state (8.28) one gets:

$$M = -A \frac{N}{V_0 \left(1 + \alpha \ (T - T_0) - k_T P \right)} H.$$

The constant A can be expressed in terms of the susceptibility χ_0, defined as the magnetic susceptibility at T_0 with $P = 0$:

$$\chi_0 = -A \frac{N}{V_0} \Rightarrow A = -\frac{V_0 \chi_0}{N},$$

yielding

$$M = \chi_0 \frac{H}{1 + \alpha \ (T - T_0) - k_T P}. \tag{8.29}$$

The situation is different in the case of metals because their behaviour depends on the conduction electron states. Diamagnetic metals display negative susceptibilities with the same order of magnitude as insulators above 100 K, with a weak temperature dependence, but at low temperatures quantum effects must be considered.

Another important class of diamagnetic materials are **superconductors**. These materials are perfect diamagnets exhibiting $\chi = -1$. They cannot be considered normal diamagnets as they are characterised by zero electrical resistance and only exist at low temperatures, below a critical temperature. Superconductivity must be treated as a different thermodynamic state characterised simultaneously by zero electrical resistance and $\chi = -1$.

8.5.2 PARAMAGNETIC SYSTEMS

As diamagnets, the paramagnetic systems only exhibit a non-zero magnetisation in an applied magnetic field. In this case, the magnetisation is proportional and parallel to the magnetic field with the positive magnetic susceptibility having typical values ranging between 10^{-3} and 10^{-6} in SI units. Paramagnetic materials are composed of atoms or molecules that present permanent magnetic moments $\bar{m}$. In zero magnetic field, all magnetic moment orientations are equally probable, and equally represented in the N particle system, so that their average sum yields a zero total magnetic moment. Under an applied magnetic field, the magnetic moments tend to align with H, the field alignment competing with the thermal misalignment that privileges a high entropy state. This competition is associated with a strong dependence of the magnetisation with temperature:

$$M = \chi \ H = \frac{C}{T} H. \tag{8.30}$$

The relation above is named **Curie law** and the constant $C = \mu_0 \frac{N}{V} \frac{m^2}{3k_B}$, designated Curie Constant, depends on N/V and on the magnetic moment m. Replacing C in (8.30),

$$M = \mu_0 \frac{m^2}{3k_B} \frac{N}{V} \frac{1}{T} H. \tag{8.31}$$

Like in the case of diamagnets, the volume depends weakly on the temperature and pressure, but due to the strong dependence of the magnetisation on temperature, the variation of the volume with temperature and pressure can be ignored in a first approximation. Note that expressions (8.30) and (8.31) diverge as T approaches 0 K. Since the system's magnetisation cannot grow indefinitely, these expressions are only valid in the high temperature range, defined as $mH/(k_B T) \ll 1$. The magnetic moments of atoms and molecules are typically of the order of 3×10^{-23} A m^2 and magnetic fields in laboratories rarely exceed 10 T, meaning that the expression is valid for $T > 10$ K for most materials that remain paramagnetic at lower temperatures. In some cases, like for some paramagnetic salts its validity extends below 1 K range. Most magnetic materials are paramagnetic at high temperatures, and the ordered state appears below a critical temperature θ. In those cases, the expression (8.30) is replaced by

$$M = \chi H = \frac{C}{T - \theta} H. \tag{8.32}$$

In the case of a transition to a ferromagnetic state, $\theta > 0$, while in the case of a transition to an antiferromagnetic state, $\theta < 0$. The equation of state (8.31) is only valid for temperatures above θ, where the material can be considered paramagnetic.

8.6   ADIABATIC DEMAGNETISATION

The most efficient way to cool down a system to low temperatures is to put it in contact with a heat reservoir, generally consisting of a fluid whose boiling point is lower than the desired temperature. Of the possible cryogenic fluids, liquid helium has the lowest boiling point (4.2 K at normal atmospheric pressure, reaching 0.7–0.8 K at low pressure (Dixit, 1938); 0.7 K is therefore the lowest temperature attainable using a cryogenic fluid. In order to cool below this limit, a different process must be used to remove heat from the system. The first method available for that purpose is a magnetic cooling process that uses a magnetic system characterised by an important change in temperature when the magnetic field is varied adiabatically. The cooling or heating of a system when the applied magnetic field is varied is called **magnetocaloric effect** and the process of cooling using this effect is named **adiabatic demagnetisation**.

Think about it...
Why does the adiabatic variation of the magnetsation of a paramagnet, results in cooling?

Answer

In a solid, the atoms vibrate around their equilibrium positions in the crystal lattice. By increasing temperature the atoms will vibrate with increasing amplitude. The magnetic moments in an ideal paramagnet were considered independent in the sense that we can discard magnetic interactions between them. Nevertheless, the magnetic moments cannot be considered independent of the material atomic structure (crystal lattice), since the magnetic moments depend on the atom's electronic states, and bonding between the atoms involves the valence electronic states. This means that the magnetic moments cannot be considered non interacting with the lattice. This interaction allows magnetic energy to be converted to atomic vibrations of the atoms and vice-versa. From the thermodynamic point of view, we can think of the magnetic system being coupled to the lattice system. If both systems are not at the same temperature, they evolve to reach thermal equilibrium. For instance, a paramagnetic salt in a high external magnetic field will have a non-null magnetisation, with the magnetic moments aligned to some extent. Removing the magnetic field in an adiabatic transformation, will leave the ordered magnetic moments in a non-equilibrium state, exhibiting lower entropy than expected, and, immediately after removing the magnetic field, the magnetic moments are considered at a "lower temperature" than the lattice. The magnetic moments increase their entropy by getting energy from the lattice, and both magnetic moments and lattice reach thermal equilibrium.

The dependence of the temperature on the magnetic field in an adiabatic transformation at constant pressure is an important quantity, a **magnetocaloric coefficient**, which can be quantified using the Maxwell relation (8.18)

$$\left(\frac{\partial T}{\partial H}\right)_S = -\mu_0 \left(\frac{\partial \mathfrak{M}_\parallel}{\partial S}\right)_H. \tag{8.33}$$

For materials having low demagnetising field and assuming the linearity condition, H can be approximated by the external field H_{ext} and $\mathfrak{M}_\parallel = V\chi H$. Therefore, (8.33) can be recast as

$$\left(\frac{\partial T}{\partial H_{ext}}\right)_S = -\mu_0 \left(\frac{\partial \mathfrak{M}_\parallel}{\partial S}\right)_{H_{ext}},$$

or

$$\left(\frac{\partial T}{\partial H_{ext}}\right)_S = -\mu_0 \frac{\left(\frac{\partial \mathfrak{M}_\parallel}{\partial T}\right)_H}{\left(\frac{\partial S}{\partial T}\right)_H} = -\frac{\mu_0 T}{c_H}\left(\frac{\partial \chi}{\partial T}\right)_H H. \tag{8.34}$$

Note that the specific heat is defined per unit volume and apart from depending on temperature, in general it will also depend on the magnetic field.

Magnetic cooling was proposed independently by Peter Debye (1844–1966) in 1926 and by William F. Giauque (1895–1982) in 1927, but experimental demonstration came only in 1933 with the work of Wander J. de Haas (1878–1960), W.F. Giauque, and collaborators.

The adiabatic demagnetisation scheme is illustrated in Figure 8.3 A. The sample to be cooled is placed in thermal contact with the cold finger that is always kept in close thermal contact with **P** (a material with important magnetocaloric effect – usually a paramagnetic salt). The adiabatic demagnetisation cooling includes two

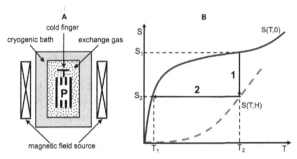

Figure 8.3   Scheme of an adiabatic demagnetisation system, with cooling being determined by the magnetocaloric material **P** (A). The processes 1 (isothermal) and 2 (adiabatic) in the (S, T) state plane showing the decrease in entropy and temperature associated with cooling by adiabatic demagnetisation (B).

thermodynamics processes indicated by 1 and 2 in Figure 8.3 B. In process 1, keeping thermal equilibrium between the paramagnetic salt **P** and a cryogenic reservoir at temperature T_1 (the thermal contact with the reservoir is achieved through exchange gas in the exchange chamber), an external magnetic field H_{ext} is applied that supplies work to the system. The magnetic moments of **P** become partially oriented in the direction of the applied magnetic field, **P** and the system's entropy decreases isothermally. In this process energy flows as heat from **P** to the cryogenic fluid. In process 2, the exchange gas is removed, and then the magnetic field is reduced to zero. In this process, magnetic work is performed by **P** that cools down to T_2 through an adiabatic process.

To calculate the temperature variation during adiabatic cooling, we will consider a paramagnetic salt as the magnetocaloric material **P**. This allows replacing the susceptibility in (8.34) using the Curie law:

$$\left(\frac{\partial T}{\partial H_{ext}}\right)_S = \frac{\mu_0 C}{c_H T} H.$$

The specific heat c_H of a paramagnetic salt has two main contributions: the lattice specific heat that increases with temperature as T^3, and the magnetic specific heat that is related with the magnetic susceptibility by $c_{Hmag} = \mu_0 \frac{H^2}{T} \chi = \mu_0 \frac{CH^2}{T^2}$. At very low temperatures, the lattice specific heat is negligible compared with the magnetic contribution, implying $c_H \approx \mu_0 \frac{CH^2}{T^2}$ for non-zero magnetic fields. Then,

$$\left(\frac{\partial T}{\partial H}\right)_S = \frac{\mu_0 C}{C \mu_0 H^2} T H \implies \frac{dT}{T} = \frac{dH}{H} \implies \ln\left(\frac{T_f}{T_i}\right) = \ln\left(\frac{H_f}{H_i}\right) \implies \frac{T_f}{T_i} = \frac{H_f}{H_i}$$

Assuming that the applied field is $\mu_0 H \sim 5$ T, the reduction of H to the magnetic field of $\mu_0 H \sim 5$ mT, will reduce the temperature from 4 K to about 4 mK. The millikelvin region is the lowest temperature range obtained with paramagnetic salts, since the assumption of a Curie law behaviour breaks down at very low temperature due to the interaction between the magnetic moments. To obtain lower temperatures, there is the possibility of using the ordering of nuclear magnetic moments, that are much smaller than the atomic ones, in a technique named **adiabatic nuclear demagnetisation**.

8.7   ABSOLUTE NEGATIVE TEMPERATURE

Thermodynamic temperature is the derivative of the internal energy with respect to entropy, or its inverse as the derivative of the entropy with respect to the internal energy, keeping all other extensive variables $\{X_i\}$ constant:

$$T = \left(\frac{\partial U}{\partial S}\right)_{\{X_i\}} \qquad \frac{1}{T} = \left(\frac{\partial S}{\partial U}\right)_{\{X_i\}}.$$

If the entropy is considered a monotonic function of energy, the conclusion is that temperature is always a positive quantity. We will now show that magnetic systems can exhibit negative absolute temperatures.

Consider a thermally isolated system composed of N localised, non-interacting magnetic moments $\vec{m}_i$, $i = 1,...,N$ with the same magnitude m, but different orientations in the presence of a uniform external magnetic field $\vec{B}$. A microstate of the magnetic system is then a microscopic configuration of the system specifying the orientation of each magnetic moment. The (magnetic) energy of the system is

$$U = -\sum_{i=1}^{N} \vec{B} \cdot \vec{m}_i$$

and the entropy S can be calculated for each value of U using Boltzmann's equation (3.40)

$$S = k_B \ln W,$$

with W representing the number of microstates that have energy U. The lowest energy state ($U = U_{min} = -NmB$) corresponds to a microstate in which all magnetic moments are aligned parallel to the magnetic field $\vec{B}$, while the highest energy state ($U = U_{max} = NmB$) corresponds to the microstate in which all magnetic moments are aligned anti-parallel to $\vec{B}$. Since for both energy states the number of possible microstates is exactly one, both energy states have the same entropy $S(U_{min}) = S(U_{max}) = 0$. The change of orientation of a single magnetic moment in each one of these two energy states corresponds to an entropy increase, and there are N possible corresponding microstates (as many as the number of magnetic moments whose orientation can be changed). However, if the initial state is U_{min}, the change of orientation of one magnetic moment corresponds to an energy increase, while if the initial state is s U_{max}, the energy will decrease when one magnetic moment changes its orientation. In the latter case $\partial U / \partial S < 0$, which corresponds to a negative absolute temperature.

For simplicity, let us consider a system comprising six magnetic moments that have only two possible orientations relative to the magnetic field, parallel or antiparallel, with $+m$ and $-m$ being the corresponding values of the component of the magnetic moment along B. The list of the possible values of U for this system, together with the number of corresponding microstates and entropy is provided in Table 8.2.

State	U	Number of microstates	S
↑↑↑↑↑↑	$-6m$	$C_0^6 = 1$	0
↓↑↑↑↑↑	$-4m$	$C_1^6 = 6$	$k_B \ln 6$
↓↓↑↑↑↑	$-2m$	$C_2^6 = 15$	$k_B \ln 15$
↓↓↓↑↑↑	0	$C_3^6 = 20$	$k_B \ln 20$
↓↓↓↓↑↑	$2m$	$C_4^6 = 15$	$k_B \ln 15$
↓↓↓↓↓↑	$4m$	$C_5^6 = 6$	$k_B \ln 6$
↓↓↓↓↓↓	$6m$	$C_6^6 = 1$	0

Table 8.2

Energy, number of microstates and entropy for six non-interacting magnetic moments in the presence of a magnetic field $\vec{B}$.

By using the values of Table 8.2 it is possible to plot the entropy as a function of the energy (Figure 8.4 A). A macroscopic system would have an almost continuous curve with the same behaviour, depicted by the full line. The $U = 0$ axis separates the region where S increases with increasing energy (positive temperatures) from the region where S decreases with energy (negative temperatures). In the region where entropy decreases with energy, the number of microstates that realise a state decreases with energy. This is only possible because the system is composed of particles that have an upper limited energy. As the system energy approaches its maximum the negative temperature increases (Figure 8.4 B), which means that the negative temperature increases with energy, but the system at negative temperature has more energy than the same system at any positive temperature. Between two systems at a negative temperature the one with a higher temperature can supply power to a system at lower temperature. In Figure 8.4 B, $T = (\partial U/\partial S) \to \infty$ at $U = 0$, which means that the temperature as a function of energy has two separate branches without any possible path between them.

Another way of looking at the relation between temperature and energy in this system is to plot the ratio $-\frac{1}{T}$ as a function of energy, where higher energy states are associated with higher $-\frac{1}{T}$ for both negative and positive temperatures (Figure 8.4 C). This plot was first suggested in 1956 by Norman Foster Ramsey (1915–2011), who also discussed the experimental observation of negative temperature systems and wrote: *The occurrence of systems at negative temperatures will necessarily be relatively infrequent since a very special combination of rarely met requirements must be satisfied before negative temperatures are even a possibility for the system.*

Negative absolute temperature appears associated with systems formed by particles whose energy is bounded from above, for high internal energies (corresponding

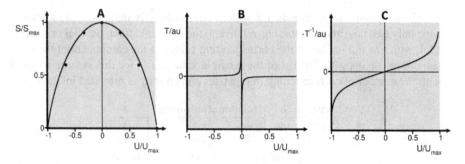

Figure 8.4   Energy as a function of entropy for six magnetic moments system characterised by limited number of energy levels (full circles), and for a similar system with a macroscopic number of magnetic moments (full line) (A). Temperature determined as the energy derivative of entropy as a function of energy (B). The symmetric of the temperature inverse orders monotonically with magnetic energy (C).

to the particles having energies in the top half part of their spectra). In this case, we speak of particle systems with **limited energy spectrum** and **population inversion**. In general, these systems are in quasi-equilibrium states because they cannot be kept completely isolated from their surroundings and they will evolve slowly to positive temperature states.

8.8   LEARNING OUTCOMES

At the end of this chapter the reader is expected to:

1. Know the fundamental equation of thermodynamics for magnetic systems.
2. Be able to define thermodynamic potentials for magnetic systems.
3. Know the magnetic coefficients and understand that they are not independent.
4. Know the equations of state for paramagnetic and diamagnetic systems.
5. Understand cooling by adiabatic demagnetisation.
6. Understand the meaning of negative absolute temperature.

8.9   WORKED PROBLEMS

PROBLEM 8.1
Consider the first process (isothermal increase of the magnetic field) in the adiabatic demagnetisation method and calculate the entropy variation of the paramagnetic salt when the magnetic field increases from zero to H.

Solution

To calculate the entropy variation, the appropriate thermodynamic potentials must be chosen for the isothermal process. Since the controlled variables are the magnetic field and temperature, the magnetic Gibbs energy is chosen:

$$dg = -sdT - \mu_0 M dH$$

with

$$-s = \left(\frac{\partial g}{\partial T}\right)_H \quad \text{and} \quad \mu_0 M = \left(\frac{\partial g}{\partial H}\right)_T .$$

For the paramagnetic material $M = \frac{C}{T}H$, and

$$\Delta g = -\mu_0 \frac{C}{T} \int_0^H H\, dH = -\mu_0 \frac{C}{T} \frac{H^2}{2} \implies g(T,H) = g(T,0) - \mu_0 \frac{C}{T} \frac{H^2}{2}$$

The entropy can be calculated as the derivative of the magnetic Gibbs energy

$$-s = \left(\frac{\partial g}{\partial T}\right)_H = -s_0 + \mu_0 \frac{C}{T^2} \frac{H^2}{2} \implies \Delta s = -\mu_0 \frac{C}{T^2} \frac{H^2}{2}$$

PROBLEM 8.2
Determine the relation between thermodynamic coefficients expressed in equation (8.27).

Solution
Use the definition of the specific heats

$$c_{M,P} - c_{H,P} = T\left(\left(\frac{\partial s}{\partial T}\right)_{P,M} - \left(\frac{\partial s}{\partial T}\right)_{P,H}\right) \tag{8.35}$$

and define the entropy per unit volume

$$s(T,H) = \frac{S(T,H)}{V} = s(T,M(T,H))$$

The derivatives appearing in equation (8.35) are related in the following manner:

$$\left(\frac{\partial s}{\partial T}\right)_H = \left(\frac{\partial s}{\partial T}\right)_M + \left(\frac{\partial s}{\partial M}\right)_T \left(\frac{\partial M}{\partial T}\right)_H$$

Multiplying the previous expression by T,

$$c_H = c_M + T\left(\frac{\partial s}{\partial M}\right)_T \alpha_H,$$

and using a Maxwell relation:

$$\left(\frac{\partial s}{\partial M}\right)_T = -\mu_0 \left(\frac{\partial H}{\partial T}\right)_M = \mu_0 \frac{\left(\frac{\partial M}{\partial T}\right)_H}{\left(\frac{\partial M}{\partial H}\right)_T} = \mu_0 \frac{\alpha_H}{\chi_T} .$$

c_H can thus be written as

$$c_H = c_M + \mu_0 T \frac{\alpha_H^2}{\chi_T} \Rightarrow c_M - c_H = -\mu_0 T \frac{\alpha_H^2}{\chi_T} .$$

And finally,

$$C_M - C_H = -\mu_0 TV \frac{\alpha_H^2}{\chi_T} .$$

8.10 SUGGESTED PROBLEMS

PROBLEM 8.3
Show that for a paramagnetic system, characterised by a Curie constant C and a heat capacity C_H the following relations hold:
a)

$$\left(\frac{\partial s}{\partial H}\right)_T = -\mu_0 C \frac{H}{T^2}\left(\frac{\partial T}{\partial H}\right)_S$$

b)

$$\left(\frac{\partial T}{\partial H}\right)_S = -\mu_0 C \frac{H}{C_H T}$$

PROBLEM 8.4
Consider a block of SiO_2 as a diamagnetic system and take the following values of magnetic susceptibility at $p = 1$ atm: -1.65×10^{-6} at $T = 20°C$ and -1.64×10^{-6} at $T = 60°C$. The system undergoes an isobaric and isothermal increase of the magnetic field from 0 to $\mu_0 H = 1.00$ T at $T = 20°C$ and $P = 1$ atm.

a) Calculate the entropy variation of the system.
b) Calculate the heat flow across the system's boundary.
c) Calculate the work done by the system.
Hint: At constant P and T equation (8.29) reduces to $\chi = \frac{1}{a+bT}$ with $a = (1 - \alpha T_0 - \kappa_T P)/\chi_0$ and $b = \frac{\alpha T_0}{\chi_0}$. a and b can be calculated from the experimental values of the susceptibility.

PROBLEM 8.5
Consider two paramagnetic systems S_1 and S_2 with the volume of $1\,cm^3$ each in a magnetic field $H = 1.0 \times 10^3$ A/m. S_1 has a magnetisation of 2.0×10^{-4} emu and S_2 has a magnetisation of 0.5×10^{-4} emu at $T = 293$ K. The corresponding heat capacities can be considered constant in the field range assumed and are equal to 0.8 J/K and 0.5 J/K respectively.

a) Calculate the susceptibility of both systems at $T = 293$ K and $T = 150$ K.
b) If the two systems are separated by a diathermal and impermeable wall, and together are isolated from the exterior, what will be the equilibrium temperature of both systems when the magnetic field is removed.

REFERENCES

1. Callen, H. B. (1960). Thermodynamics. Wiley.

2. Cooke, A. H. & Meyer, W. P. (1956) Proc. R. Soc. Lond. A 237: 395-403.

3. Debye, P. (1926) Annals of Physics 8:1154-1160.

4. Dixit, K. R. (1938) Current Science 6: 589-599.

5. Giaque, W. F. (1927) J. Am. Chem. Soc. 49:1864-1870.

6. Giaque, W. F. (1933) Phys. Rev. 43:768.

7. de Haas, W. J., Wiersma, E. C., Kramers, H. A. (1933) Physica 1:1-13.

8. Ramsey, N. F. (1956) Phys. Rev. 103: 20-28.

9 Thermal Radiation

Electromagnetic radiation interacting with matter reaches a state of thermodynamic equilibrium with a definite temperature called thermal radiation. This chapter provides a brief introduction to the study of this system, in what may seem an unexpected application of thermodynamics. Yet, thermal radiation is the simplest thermodynamic system studied in the present book.

9.1 INTRODUCTION

The application of thermodynamics to electromagnetic radiation is often presented as a brief episode in an exciting story, involving a series of experiments and theoretical contributions that ultimately lead to the development of quantum theory. Indeed, the concept of energy quantisation was a truly revolutionary outcome of the study of thermal radiation, and its importance sometimes overshadows the accomplishment of extending the scope of thermodynamics beyond the systems it was first formulated for. Until then, thermodynamics dealt with the equilibrium of such systems, and radiation, apart from its different nature, is not normally associated with equilibrium. The concept of cavity radiation, introduced by Kirchhoff, brought radiation into the realm of thermodynamics, and created a new model system, simpler even than a fluid. In this chapter, we will explore this system and see how thermodynamics, overarching the whole of Physics, anticipates quantum theory. We will then apply the thermodynamics of radiation to stars and to the universe as a whole.

9.2 KIRCHHOFF'S LAW AND BLACK BODY RADIATION

Gustaf Robert Kirchhoff (1824–1887) is well known for his electric circuits laws, but he was also a prominent figure in the field of spectroscopy, whose work set the ground for spectrochemical analysis in laboratory and in astrophysics. Together with Robert Wilhelm Bunsen (1811–1899), he undertook a systematic analysis of the light emitted by incandescent substances and discovered the unique spectral patterns that characterise each element. They also realised that light viewed through a layer of gas of a given element would loose the same frequencies the hot element emits, and went on to correctly deduce that dark lines in the solar spectrum are caused by absorption by chemical elements in the solar atmosphere.

His experimental work on emission and absorption spectra lead Kirchhoff to consider the relation between these two properties in an equilibrium setting. He thought of electromagnetic radiation enclosed in a box with opaque walls maintained at uniform temperature T. Let $E_w(\lambda, T)$ denote the **spectral emissive power** of the cavity walls, defined as the energy emitted in the wavelength range $(\lambda, \lambda + d\lambda)$ per unit time and unit surface area. We know that radiation, visible light in particular, interacts differently with different materials, and the index w highlights the dependence of this quantity on the material the walls are made of. $E_w(\lambda, T)$ will depend on the

DOI: 10.1201/9781003091929-9

nature of the cavity walls, and so will the **spectral absorptivity** $A_w(\lambda, T)$, defined as the fraction of the **spectral intensity** (the incident energy in the wavelength range $(\lambda, \lambda + d\lambda)$ per unit time and unit surface area) absorbed by the wall.

In contrast, the quality of the radiation inside the cavity cannot depend on the properties of the walls and must depend only on their temperature. Imagine it was not so and assume that we have two such cavities at temperature T connected by a very narrow tube through which radiation may pass. If there is any difference between the energy carried by radiation on each side, it would be possible to transfer a finite amount of energy from side A to side B, increasing the temperature of B in the process. We would then have a flow of energy from a colder to a hotter body with no other change, in violation of the second law. By adding a colour filter to the imaginary narrow tube in the preceding argument we conclude that the spectral intensity of the cavity radiation must also depend on temperature alone, and not on any other physical property of the enclosure.

In equilibrium, there is no net energy flow in the walls, and therefore

$$E_w(\lambda, T) = A_w(\lambda, T)I(\lambda, T), \tag{9.1}$$

where $I(\lambda, T)$ is the **universal function** that characterises the spectral intensity of thermal radiation. Equation (9.1) is **Kirchhoff's law** of thermal radiation, and its consequences can be explored in different directions.

On one hand, it means that emissive power and absorptivity are simply related, for all bodies and in all frequency ranges. Dark surfaces, being good absorbers, must also be good emitters in thermal equilibrium. Assume this was not true, and consider two objects, one dark and another one light, with the same emissive power, enclosed in a cavity in thermal equilibrium. Then, either the dark object will absorb more energy than it emits, and heat up, or the light object will absorb less energy than it emits, and cool down (or both). In either case, the process would entail a flow of energy from a colder to a hotter body with no other change. Not surprisingly, we find that contradicting this consequence of (9.1) leads to a violation of the second law, by an argument similar to the one used above to derive it.

The deeper meaning of Kirchhoff's law stems from the construction of the universal function $I(\lambda, T)$ that establishes cavity radiation as a simple thermodynamic system whose properties can be derived using the theory. To overcome the difficulty that cavity radiation cannot really be studied experimentally, Kirchhoff went one step further in abstraction and defined a perfect **black body** as one for which $A(\lambda, T) \equiv 1$ and so $E(\lambda, T) = I(\lambda, T)$. So, by definition, a black body absorbs all incident radiation, reflecting none, and its emissive power $E(\lambda, T)$ coincides with the spectral intensity of thermal radiation. In spite of its idealised nature, a black body can actually be built, by punching a small hole in the walls of an otherwise closed cavity with rough or irregular walls (Figure 9.1). If the hole is small enough, almost all incident radiation will remain inside the cavity as it is partially reflected by the walls. The hole will therefore behave as a perfect black body, and the radiation it emits in thermal equilibrium can be used to probe cavity radiation. Indeed, not long after Kirchhoff's challenge this construction was being used to conduct experiments on what became known in the literature as black body radiation.

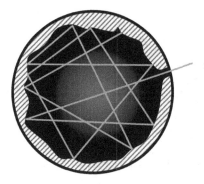

Figure 9.1   An approximate realisation of an ideal black body.

9.3   THERMODYNAMICS OF RADIATION AND THE STEFAN-BOLTZMANN LAW

We now have everything we need to approach the study of a new model, a cavity with radiation instead of a vessel containing a fluid. The spectral intensity of thermal radiation $I(\lambda, T)$ contains all the information about the system in equilibrium at temperature T. Instead of $I(\lambda, T)$, the spectral energy density $u(\lambda, T)$, defined as the the energy per volume contained in the wavelength range $(\lambda, \lambda + d\lambda)$, is better suited for thermodynamics. The two functions are related simply by $I(\lambda, T) = \frac{c}{4} u(\lambda, T)$, which holds for any quantity propagating isotropically at velocity c (the factor of $1/4$ comes from the isotropy property).

LAMBERT COSINE RULE

If we consider an energy density $u(\lambda, T)$ propagating at speed c in a given direction, we obtain radiation intensity $c\, u(\lambda, T)$ on a surface placed perpendicularly to the propagation direction. If instead the direction of propagation makes an angle θ with the normal $\mathbf{n}$ to the surface, then the radiation intensity on the surface will be $c\, u(\lambda, T) \cos \theta$ (Figure 9.2). In order to take into account all incoming directions, we must then average $c\, u(\lambda, T) \cos \theta$ over all possible solid angles Ω. Using spherical coordinates (θ, φ) with respect to the normal to the surface,

$$I(\lambda, T) = c\, u(\lambda, T) \int_{\Omega} \cos \theta \, \tfrac{d\Omega}{4\pi} = \tfrac{c\, u(\lambda, T)}{4\pi} \int_0^{2\pi} \int_0^{\pi/2} \cos \theta \sin \theta \, d\theta \, d\varphi = \tfrac{c\, u(\lambda, T)}{4}$$

Note that the integral in θ in taken only between 0 and $\pi/2$ because only incoming directions contribute to the intensity.

We have then that the overall energy density is a function of T only,

$$u(T) = \int_{\lambda} u(\lambda, T) \, d\lambda, \tag{9.2}$$

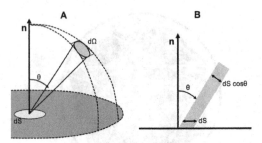

Figure 9.2   Geometry of Lambert's cosine rule.

and therefore the energy of the system is $U = Vu(T)$. In order to learn more about the function $u(T)$, let us consider an enclosure with a piston, so that the radiation "gas" can expand or contract, and write down the fundamental equation in the form

$$T dS = dU + P dV, \tag{9.3}$$

so that

$$T \left(\frac{\partial S}{\partial V} \right)_T = \left(\frac{\partial U}{\partial V} \right)_T + P. \tag{9.4}$$

Using $U = Vu(T)$ and Maxwell's relation (5.26)

$$\left(\frac{\partial S}{\partial V} \right)_T = \left(\frac{\partial P}{\partial T} \right)_V, \tag{9.5}$$

this can be recast as

$$T \left(\frac{\partial P}{\partial T} \right)_V = u(T) + P. \tag{9.6}$$

In the scope of classical electrodynamics, the radiation pressure P had been shown to be related with the energy density u by $P = u/3$. This relation is exact and can be derived as well in the framework of a quantum theory of radiation. Introducing in (9.6), we obtain a simple differential equation for the unknown function $u(T)$:

$$\frac{T}{3} \frac{du}{dT} = \frac{4}{3} u(T), \quad \frac{d}{dT} \log u = \frac{4}{T}, \tag{9.7}$$

yielding

$$u(T) = a \, T^4, \tag{9.8}$$

where a is an integration constant, called the radiation constant. In terms of the overall intensity of thermal radiation $I(T) = \int_\lambda I(\lambda, T) \, d\lambda$ we have then

$$I(T) = \frac{a c}{4} T^4 \equiv \sigma T^4, \tag{9.9}$$

where σ is called the Stefan constant, because (9.9) was established in 1879 by Josef Stefan (1835–1893) as an approximate empirical relation, based on experimental results on the emissive power of hot metals. The thermodynamic derivation of (9.9) we just revisited was given by Boltzmann in 1884, and (9.9) became known as the **Stefan-Boltzmann law.** It was not until 1897 that careful experiments by Lummer and Pringsheim on black body radiation showed that the law is correct with high precision.

Unlike the Stefan-Boltzmann law, the value of the Stefan constant cannot be derived in the scope of classical physics. It can be defined exactly in terms of other fundamental constants, including Planck's constant h, as

$$\sigma = \frac{2\pi^5 k_B^4}{15 h^3 c^2} \approx 5.6704 \ 10^{-8} \ \text{W m}^{-2} \ \text{K}^{-4}. \tag{9.10}$$

The thermodynamic relations we have established until now for radiation are the equations of state

$$U = a V T^4, \tag{9.11}$$

and

$$P = \frac{a}{3} T^4. \tag{9.12}$$

From (9.11) we obtain the heat capacity

$$C_V = \left(\frac{\partial U}{\partial T} \right)_V = 4 a V T^3, \tag{9.13}$$

with $C_V \to 0$ as $T \to 0$, in agreement with the third law. We also know from (3.30) that $C_V = T \left(\frac{\partial S}{\partial T} \right)_V$, so we may integrate C_V / T given by (9.13) to obtain the entropy of the system as a function of T and V:

$$S = \frac{4}{3} a V T^3, \tag{9.14}$$

where we have set to zero the value of the entropy at $T = 0$. Combining (9.11), (9.12), and (9.14) we obtain different alternative expressions for the internal energy,

$$U = a V T^4 = \frac{3}{4} T S = 3 \, PV, \tag{9.15}$$

from which the other thermodynamic potentials follow :

$$F = U - TS = -\frac{1}{4} T S = -\frac{1}{3} a V T^4 \tag{9.16}$$

$$H = U + PV = TS = \frac{4}{3} a V T^4 \tag{9.17}$$

$$G = H - TS \equiv 0 \tag{9.18}$$

The fact that the Gibbs potential G comes out identically zero for the radiation "fluid" deserves some attention. First, it means that, in accordance with equation (9.16) for the free energy F, no forms of work other than expansion work are possible in this system. If we think of the radiation "gas" as a system of N particles, then we have

$G = \mu N \equiv 0$, and so the chemical potential of the system $\mu \equiv 0$. But by looking closer at the thermodynamic relations we derived, we see that the equation of state (9.12) implies that P and T are not independent variables, and therefore neither G nor μ are well defined for this system. We know in general that intensive variables are conjugate to conserved extensive variables, so our observation has a deep meaning: the quantity particle number that came up naturally when we studied the ideal gas has no analogue for the radiation fluid, whose "particles" would not be conserved in a closed system. Indeed, the quantum mechanical description of the system is that of a photon gas in thermal equilibrium, and photons can be emitted from or absorbed by the cavity walls. Although it makes sense in that description to compute the average number of photons in the cavity, and photons share some properties of classical particles, their thermodynamic status is clearly that of energy, not particles.

Table 9.1 sums up the differences between our two main model systems, the ideal gas and the photon "gas",

	Ideal Gas	**Photon Gas**
Internal energy	$U = \frac{3}{2}Nk_BT$	$U = a\,VT^4$
Volume dependence	$\frac{\partial U}{\partial V} = 0$	$\frac{\partial u}{\partial V} = 0$
Pressure	$P = Nk_BT/V = \frac{2}{3}u$	$P = \frac{1}{3}aT^4 = \frac{1}{3}u$
Heat capacity	$C_V = \frac{3}{2}Nk_B$	$C_V = 4aVT^3$
Entropy	$S = Nk_B\left(\log\frac{V}{N\lambda_T^3} + \frac{5}{2}\right)$	$S = \frac{4}{3}aVT^3$
Adiabatics	$TV^{\gamma-1} = constant$	$TV^{1/3} = constant$
Chemical potential	$\mu = -k_BT\log\frac{V}{N\lambda_T^3}$	$\mu = 0$

Table 9.1

Comparison of the thermodynamic properties of the two model systems.

9.4 WIEN'S DISPLACEMENT LAW AND THE BLACK BODY SPECTRUM

Let us go back to the spectral energy density $u(\lambda, T)$. By the end of the nineteenth century, the determination of Kirchhoff's universal function was one important open problem in Physics, and experimental results were becoming available. As we saw in the previous section, the complete understanding of thermal radiation involves energy quantisation, but it was still possible, following Boltzmann's derivation, to go further into the derivation of the black body spectrum using only classical thermodynamics. This is what Wilhelm Wien (1864–1928) achieved in 1893, a land-mark contribution for which he received the Nobel Prize for Physics in 1911. As in Boltzmann's derivation of (9.9), Wien's argument combines thermodynamics with Kirchhoff's law.

The starting point will be to consider the adiabatic expansion of the radiation "gas". Since $dQ = dU + PdV = 0$, we have

$$d(uV) + \frac{1}{3}udV = 0$$

$$Vdu + udV + \frac{1}{3}udV = 0$$

$$\frac{du}{u} = -\frac{4}{3}\frac{dV}{V}$$

and, integrating, $u = constant\ V^{-\frac{4}{3}}$. But $u = aT^4$, so, in accordance with (9.14),

$$TV^{\frac{1}{3}} = constant$$

or, in terms of the linear size r of the expanding volume V,

$$T \sim \frac{1}{r}. \tag{9.19}$$

Next comes a geometrical argument to show that, if we follow a particular set of waves inside the expanding cavity, their wavelength λ will change as $\lambda \sim r$ due to reflections in moving wall. Consider a plane mirror moving to the right with velocity u as shown in Figure 9.3 and the reflection of two wave fronts of an incoming wave separated by a full period Δt. Since during that time interval the wall receded $\Delta X = u\ \Delta t$, the second wave front will see its path increase and therefore its wavelength increase by the same amount, say $\Delta\lambda$. Indeed, since we are comparing two in-phase consecutive wave fronts, the distance between them is the wavelength, and the figure tells us that $\Delta\lambda = \overline{AB} + \overline{BN} = \overline{A'N} = 2\Delta X \cos\theta$, θ being the incidence angle. Then,

$$\Delta\lambda = 2\cos\theta\ u\ \Delta t \approx 2\cos\theta\ u\ \frac{\lambda}{c} \tag{9.20}$$

is the wavelength change associated with one reflection. Consider now the reflecting walls of an expanding, spherical volume to compute the change in wavelength associated with a small change Δr in the radius of the sphere. Since the incidence angle θ will be approximately the same in consecutive reflections, we only have to take into account how many reflections in the expanding walls there will be (Figure 8.3 B). The time between consecutive reflections being $2\ r\cos\theta/c$, the number of reflections in a small time interval $\Delta r/u$ will be $c\ \Delta r/(2\ u\ r\cos\theta)$. Hence, from (9.20) the overall wavelength shift associated with an increase Δr in the radius of the sphere is

$$\Delta\lambda = \lambda\frac{\Delta r}{r}$$

A

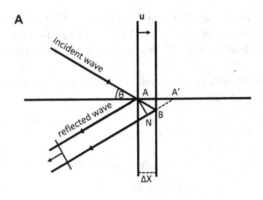

B

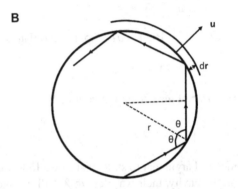

Figure 9.3   Change in wavelength due to a single reflection on a moving plane (A). Consecutive reflections on the inner surface of an expanding sphere (B).

and, integrating,

$$\lambda \sim r. \tag{9.21}$$

Combining (9.19) and (9.21), we get

$$T \sim \lambda^{-1}. \tag{9.22}$$

This is true in particular if we follow the set of waves that correspond to the maximum of the spectral energy density, $u(\lambda_{peak}, T)$, as the equilibrium temperature slowly changes through adiabatic expansion or compression of the cavity. Hence, (9.22) implies that the maximum of the black body radiation spectrum is attained at a wavelength that shifts with temperature according to

$$\lambda_{peak} T = b, \tag{9.23}$$

where the constant b, called Wien's constant, has the value $2.898 \ 10^{-3}$ m K. Equation (9.23), known as **Wien's displacement law**, was found to be in very good agreement with experiment. Qualitatively, it implies that hotter sources will emit bluer light, or higher frequency electromagnetic radiation, since λ_{peak} is the dominant colour of the spectrum. This is related with common everyday experiences, such as body temperature determination by infra-red camera imaging, or the change in colour of metal heated at high temperatures (Figure 9.4).

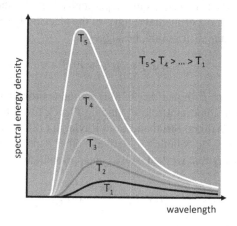

Figure 9.4   Black body radiation becomes bluer, as well as brighter, as the temperature increases.

Wien's contribution goes beyond the displacement law. His goal was the determination of the universal function $u(\lambda, T)$, so he combined (9.22) with Stefan-Boltzmann (9.8) for the adiabatic expansion of the cavity from an initial temperature T_1 to a final temperature T_2. At all stages of the expansion, the radiation will be in thermal equilibrium and therefore will maintain the black body spectral and overall energy density, so the function $u(\lambda, T)$ must scale appropriately. Considering the energy density in an arbitrary wavelength interval, it will change with T^4 to comply with the Stefan-Boltzmann law as its wavelength is shifted according to Wien's displacement law. Hence,

$$\frac{u(\lambda_1, T_1)d\lambda_1}{u(\lambda_2, T_2)d\lambda_2} = \frac{T_1^4}{T_2^4} \tag{9.24}$$

must hold with $\lambda_1 T_1 = \lambda_2 T_2$. Introducing this relation, (9.24) becomes

$$\frac{u(\lambda_1, T_1)}{u(\lambda_2, T_2)} = \frac{T_1^5}{T_2^5},$$

or

$$u(\lambda, T)\lambda^5 = constant. \tag{9.25}$$

Any function of the product λT may play the role of the constant in the second member, so the most general form for the equilibrium spectral energy density is

$$u(\lambda, T) = \lambda^{-5} f(\lambda T). \tag{9.26}$$

In terms of the frequency $v = c/\lambda$, Wien's ansatz for the energy density is

$$u(v,T) = v^3 f(v/T), \tag{9.27}$$

where again f is an arbitrary function (see problem 8.3).

Wien went still further to conjecture a particular form for the unknown function f. Taking into account experimental results for the black body spectrum and inspired by Maxwell-Boltzmann distribution, he suggested

$$u(v,T) = \alpha \, v^3 e^{-\beta v/T}, \tag{9.28}$$

with some constants α, β. Equation (9.28) is a good approximation at high frequencies and it played an important role in guiding Planck towards the correct law. Indeed, in 1900, John William Strutt and Lord Rayleigh (1842–1919), used a classical argument and wave mode counting to derive another expression for the black body spectral energy density that worked well as an approximation at low frequencies,

$$u(v,T) = \frac{8\pi v^2}{c^3} k_B T, \tag{9.29}$$

and also verified Wien's ansatz. Equation (9.29) is known as the **Rayleigh-Jeans law**.

The correct result is Planck's radiation law,

$$u(v,T) = \frac{8\pi h v^3}{c^3} \frac{1}{e^{hv/k_B T} - 1}, \tag{9.30}$$

published on that same year. The comparison in Figure 9.5 shows how it interpolates between Wien's law, at high frequencies, and Rayleigh-Jeans law, at low frequencies. However, Planck's law is much more than a heuristic interpolation of (9.28) and (9.29). It introduces a single new constant, Planck's constant h, whose appropriately chosen value yields a superb agreement with the experimental results. In deriving it, Planck follows a convoluted line of reasoning, based on a model of cavity where matter is represented by a system of electrically charged oscillators of all frequencies that exchange energy with the electromagnetic field. In thermodynamic equilibrium, the energy density spectrum for this particular system must also be given by Kirchhoff's universal function. In the study of his chosen model system, Planck departs from purely classical arguments only to impose the form of the function $S(E)$ for the entropy of an oscillator that leads to the simplest interpolation of Wien's approximation and Rayleigh-Jeans law. To justify this assumption, Planck postulated that the oscillators can emit and absorb electromagnetic radiation only in finite amounts of energy of size hv.

This is the revolutionary ingredient of (9.30), and Planck won the Nobel prize in Physics 1918 "for his discovery of energy quanta", but he did not actually propose that electromagnetic radiation is quantised, and even struggled for many years to come to terms with the full consequences of his own contribution. It was Albert Einstein (1879–1955) who first thought of light quanta, or photons, as real physical

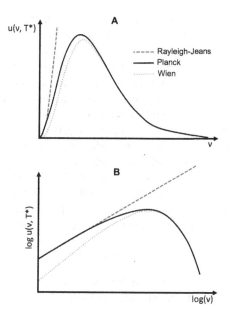

Figure 9.5   Comparison at a given temperature T^* of Planck's radiation law with Wien's and Rayleigh-Jeans in linear scale (A) and log scale (B).

objects whose existence explained not only Planck's law but also the photoelectric effect and other phenomena involving the interaction of radiation with matter that were being discovered at the time. By calling our universal model system a photon gas in the previous section we were anticipating the status it would get in the framework of quantum theory. But it is remarkable how far into this framework mere thermodynamics, together with experimental results on black body radiation, has lead physicists and Physics.

9.5   THERMAL RADIATION AND ASTROPHYSICS

The physics of thermal radiation laid the foundations for the development of twentieth century astrophysics. The connections between the two fields go back to the work of Samuel Langley (1834–1906), an astrophysicist and aeronautical pioneer. In a paper published in 1886, Langley reported a displacement towards smaller wavelengths with increasing temperature in the radiation emitted by heated copper, anticipating Wien's displacement law. Langley was also the inventor of the bolometer, an instrument to measure electromagnetic radiation, initially in the infrared part of the spectrum. Further developed by Lummer and coworkers in Germany, this instrument allowed the very precise measurements of thermal radiation that accompanied the theoretical developments covered in the preceding sections.

It also began to be used in astrophysics, to determine the temperature and the luminosity – the total emitted electromagnetic power – of the sun and of some of the stars. Although stars are not perfect black bodies, their spectrum can be fitted quite well to a Planck curve (9.30) at a given temperature, overlaid by a few dips that correspond to absortion lines of the stellar atmosphere. A precise determination of the temperature of stars can be achieved by comparing the energy emitted in different narrow frequency or wavelength bands to obtain a colour index from which the temperature can be determined directly. This relation can be deduced from Planck's law, and can be improved to include information on how different types of stars deviate from black body behaviour, as well as other corrections. The method yields for the Sun an intermediate value of the colour index, and a surface temperature of 5800 K. Two stars of the Orion constellation, Bellatrix and Betelgeuse, are examples of extreme surface temperature values, with 22,000 K and 3500 K, respectively.

Bluer, hotter stars are also brighter, according to Stefan-Boltzmann law (9.9). More precisely, the emitted power per unit area of the surface of the star is approximately given by σT^4, and therefore the luminosity L will depend only on temperature T and size according to

$$L = 4\pi\sigma r^2 T^4, \tag{9.31}$$

where r is the star's radius. Notice that luminosity is an intrinsic property of the star, in contrast with its apparent magnitude, which depends also on the star's distance to the Earth.

As astronomers developed methods to determine these distances, it became possible to measure L for a large set of stars, whose temperature was also known through their colour index. A temperature-luminosity scatterplot, called a Hertsprung-Russell diagram (HRD, for short) in honour of the astronomers who first thought of presenting in this way the information available at the beginning of the twentieth century, became a centrepiece in the development of stellar physics. It showed that stars grouped in certain regions of the diagram forming well defined families, and guided the first theories of stellar structure and evolution. More than a century later, astrophysicists are still adding information and detail to the HRD and using it to learn about the physics of stars.

But the most spectacular manifestation of the laws of black body radiation in astrophysics was also the least expected, and it came later. Two American radio astronomers, Arno Penzias and Robert Wilson, discovered accidentally in 1965 a faint but pervasive, highly uniform and isotropic signal, with a black body spectrum at a temperature of around 3 K. This low temperature thermal **cosmic microwave background radiation** (CMB) had been predicted as a remnant of the first stages of the creation of the universe by proponents of the Big Bang theory, but at the time of Penzias and Wilson discovery this was not yet the prevailing cosmological model. Indeed, even the fact that the universe is expanding was not uncontroversial at that time, and the discovery of the CMB became the most important single contribution in support of our current picture of the universe and its evolution. For that reason, Penzias and Wilson together won one half of the 1978 Nobel Prize in Physics for their serendipitous discovery.

Let us consider the physics of this space filling radiation as a photon gas in adiabatic expansion as the universe expands. The corresponding energy density u computed from Stefan-Boltzmann law for the currently observed temperature yields a very small value, when compared with the contribution of ordinary matter. Denoting by $L = V^{1/3}$ the length scale of the expanding universe, the thermodynamics summarised in Table 9.1 tells us that the universe cools down as it expands with $T \sim L^{-1}$ and $u \sim T^4 \sim L^{-4}$, meaning that temperature and energy density of radiation are unbounded and must have been huge in the early universe – it all started with a Hot Big Bang. Equally important is the fact that the product $\lambda\,T$ is kept constant in this expansion, because radiation must be redshifted as $\lambda \sim L$, λ being a length. Together these imply that the black body form of the spectrum (9.30) is preserved by the expansion, so that, provided there were at some stage enough interactions for thermal equilibrium to be established, in the absence of further interactions the spectrum will always be thermal from then on.

But how did all the radiation in the universe thermalise in the first place? When the universe was much smaller, and the temperature much higher, photons were constantly interacting with matter. Hydrogen atoms were unstable because typical photon energies were larger than the ionisation energy, and so the universe was a hot, opaque plasma of free electrons and protons. As the universe expands and cools down, this ionisation process becomes progressively less likely, until at about $T = 3000$ K there are no longer enough energetic photons to interact with matter. At this stage, called the recombination stage, stable hydrogen atoms form and, in a relatively short time scale, the universe becomes transparent. This phenomenon is called photon decoupling, because from then on, the radiation component evolved independently, cooling and redshifting while preserving its thermal equilibrium spectrum, until it reached the current temperature of 2.72548 ± 0.00057 K. Since $T \sim L^{-1}$, decoupling occurred when the universe was about one thousand times smaller than its present size,

The CMB continues to be the central piece of modern cosmology and several large scale missions have been devoted to measuring it as precisely as possible. The tiny variations in temperature of the radiation coming from different directions have been explained in the scope of the standard cosmological model and are among the main leads available to develop our understanding of the evolving universe.

9.6   LEARNING OUTCOMES

At the end of this chapter the reader is expected to:

1. Understand the concept of cavity radiation, and how how it paves the way to study electromagnetic radiation as a thermodynamic system.
2. Apply the theory developed in Part I to thermal radiation to derive the law of Stefan-Boltzmann and Wien's displacement law.
3. Learn that, in contrast, Planck's law cannot be obtained without taking into account the quantum nature of electromagnetic radiation.
4. Know about the main applications of thermal radiation in astrophysics.

9.7   WORKED PROBLEMS

PROBLEM 9.1
Star temperatures shown in HRD diagrams correspond to the star's outer layers, where the measured black body like radiation comes from. Towards the centre, the temperature increases steeply, attaining several million kelvin at the core. There is a steady radial outflow of energy and a stable temperature gradient, where radiative heat transfer, as well as other mechanisms, play a role. Consider a simple model of a core burning nuclear fuel at a temperature T, surrounded by a thin dust cloud heated by the core. Show that, in radiative equilibrium, the dust cloud reduces by half the outflow of energy, and its temperature is $T' = 2^{-1/4}T$ (adapted from [1]).

Solution

Let p be the power emitted by the core, and p'_o, p'_i the power emitted by the outer surface and by the inner surface of the dust layer. We are assuming that the layer is thin, and so $p'_o = p'_i = p'$. In radiative equilibrium, the total power outflow from the dust cloud, $2p'$, is balanced by the power inflow p coming from the core. Thus, $p' = p/2$ – the power radiated to the outside is reduced by half. According to Stefan-Boltzmann law, $p'/p = (T'/T)^4 = 1/2$, so $T' = 2^{-1/4}T$.

9.8   SUGGESTED PROBLEMS

PROBLEM 9.2
What is C_p for cavity radiation?

Hint: Consider $C_P = \left(\frac{\partial H}{\partial T}\right)_P$.

PROBLEM 9.3
Using the relations $\lambda = c/v$, $d\lambda = -\frac{c}{v^2}dv$, derive (9.27) from (9.26).

PROBLEM 9.4
What's the wavelength of maximum emission associated with the blackbody temperature of the sun, $T = 5800$ K ? In what region of the electromagnetic spectrum does it fall?

PROBLEM 9.5
The law of Stefan-Boltzmann can be used to relate the temperature of an orbiting planet with that of its star. Treating the Sun and the Earth as black bodies, show that the ratio T_E/T_S of the Earth's temperature to that of the Sun is given by $\sqrt{R_S/(2L)}$, where R_S is the Sun's radius and L is the distance between the Sun and the Earth.

PROBLEM 9.6

The Sun's surface temperature is T_o = 5770 K and the total power it emits is P_o = 3.826 10^{26} W. Another star is observed with a surface temperature of 4400 K and a total emitted power of 0.22 P_o. Find the radius of that star.

PROBLEM 9.7

The CMB has a black body spectrum at a temperature of T = 2.725 K. Find the peak frequency and the corresponding wavelength. Compute the total energy density of this microwave background (adapted from [2]).

REFERENCES

1. Huang, K. (2001) Introduction to Statistical Physics. CRC Press.

2. Liddle, A. (2020) An Introduction to Modern Cosmology. Wiley.

3. Longair, M. (2003) Theoretical Concepts in Physics. Cambridge University Press.

4. Luscombe, J. (2020) Thermodynamics. CRC Press.

5. Mehra, J. (2001) The Golden Age of Theoretical Physics. World Scientific.

Index

Note: Locators in *italics* represent figures in the text.

Printed in the United States
by Baker & Taylor Publisher Services